# Building Data-Driven Applications with Danfo.js

A practical guide to data analysis and machine learning using JavaScript

**Rising Odegua**

**Stephen Oni**

DANFOPRESS IS AN IMPRINT OF PACKT PUBLISHING

# Building Data-Driven Applications with Danfo.js

Copyright © 2021 Packt Publishing

**Group Product Manager**: Kunal Parikh
**Publishing Product Manager**: Aditi Gour
**Senior Editor**: Roshan Kumar
**Content Development Editor**: Tazeen Shaikh
**Technical Editor**: Sonam Pandey
**Copy Editor**: Safis Editing
**Project Coordinator**: Aparna Ravikumar Nair
**Proofreader**: Safis Editing
**Indexer**: Rakhi Nair
**Production Designer**: Roshan Kawale

First published: September 2021

Production reference: 1190821

Published by Packt Publishing Ltd.
Livery Place
35 Livery Street
Birmingham
B3 2PB, UK.

ISBN 978-1-80107-085-0

www.packt.com

*To the memory of my father, Odegua, for all the sacrifices and dedication to ensure I have a good life. To my mother, Success Odegua, thank you for all the support. To my siblings, Kelvin, Harrison, Peace, Efosa, and Jerimiah, I love you all.*

*– Rising Odegua*


*I dedicate this to my parents (Mr and Mrs Oni) and to my lovely younger sister, and also to my friends and everyone who supported me during this period.*

*– Stephen Oni*

# Contributors

## About the authors

**Rising Odegua** is the co-creator of Danfo.js and Dnotebook. He is a software and machine learning engineer with experience building data-driven applications in languages such as Python and JavaScript. Rising is actively engaged in building the developer ecosystem by giving talks, providing direct/indirect mentorship, working on open source projects, hosting meetups, and writing tutorials.

**Stephen Oni** is the co-creator of Danfo.js and Dnotebook. He is a software developer at Datopian, building data-driven platforms and tools. He is also an open source developer building tools that intersect the web and machine learning.

## About the reviewers

**Gant Laborde** is an owner of Infinite Red, mentor, adjunct professor, published author, and award-winning speaker. For 20 years, he has been involved in software development and is still going strong today. He is recognized as a Google Developers Expert in web and machine learning, but informally he is an *open sourcerer* and aspires to one day become a mad scientist. He blogs, videos, and maintains popular repositories for the community. He is CIO and owner of Infinite Red, a React Native and AI consulting company. He is also the author of *Learning TensorFlow.js (O'Reilly Media, Inc.)*. Special thanks to the TensorFlow.js community for being welcoming, and to my loving family for their support.

**Matvii Hodovaniuk** is a full-stack developer. He likes to work with JavaScript tools and explore new technologies. He works on writing skills and tools that can help him with writing both code and text during his free time.

# Table of Contents

**Preface**

## Section 1: The Basics

## 1

**An Overview of Modern JavaScript**

| | | | |
|---|---|---|---|
| Technical requirements | 4 | Understanding the this property | 23 |
| Understanding the difference between let and var | 4 | Arrow functions | 25 |
| | | Promises and async/await | 27 |
| var allows the redeclaration of variables | 5 | Cleaning callbacks with promises | 29 |
| var is not a blocked scope | 5 | async/await | 32 |
| Destructuring | 7 | Object-oriented programming and JavaScript classes | 34 |
| Spread syntax | 8 | | |
| Spreading or unpacking an iterable into an array | 8 | Classes | 35 |
| Creating new objects from existing ones | 9 | Inheritance | 37 |
| Function arguments | 10 | Setting up a modern JavaScript environment with transpilers | 40 |
| Overview of scopes and closures | 10 | Babel | 41 |
| Scope | 11 | Webpack | 41 |
| Closure | 12 | | |
| Understanding Array and Object methods | 14 | Unit testing with Mocha and Chai | 48 |
| Array methods | 14 | Setting up a test environment | 49 |
| Objects | 18 | Summary | 54 |

# Section 2: Data Analysis and Manipulation with Danfo.js and Dnotebook

## 2

## Dnotebook - An Interactive Computing Environment for JavaScript

| | | | |
|---|---|---|---|
| Technical requirements | 60 | Loading external packages | 68 |
| Introduction to Dnotebook | 60 | Loading CSV files | 71 |
| Setup and installation of | | Getting a div container for plots | 72 |
| Dnotebook | 62 | Gotchas when using a for loop | 73 |
| Basic concepts behind | | Working with Markdown cells | 74 |
| interactive computing | | Creating a Markdown cell | 75 |
| in Dnotebook | 63 | Adding images | 75 |
| Cells | 63 | Headings | 76 |
| Code cells | 64 | Lists | 77 |
| Markdown cells | 65 | | |
| Persistence/state | 66 | Saving notebooks | 78 |
| Writing interactive code | 68 | Summary | 79 |

## 3

## Getting Started with Danfo.js

| | | | |
|---|---|---|---|
| Technical requirements | 82 | Sorting | 105 |
| Why you need Danfo.js | 82 | Filtering | 108 |
| Installing Danfo.js | 84 | Arithmetic operations | 110 |
| Introducing Series and | | Logical operations | 114 |
| DataFrames | 85 | Data loading and working with | |
| Series | 85 | different | |
| DataFrames | 93 | file formats | 117 |
| Essential functions and | | Transforming a DataFrame into | |
| methods in Danfo.js | 101 | another file format | 120 |
| loc and iloc indexing | 101 | Summary | 121 |

# 4

# Data Analysis, Wrangling, and Transformation

| | | | |
|---|---|---|---|
| Technical requirements | 124 | Encoding DataFrames and Series | 142 |
| Transforming data | 124 | Combining datasets | 146 |
| Replacing missing values | 125 | DataFrame merge | 147 |
| Removing duplicates | 129 | Data concatenation | 153 |
| Data transformation with the map function | 131 | Series data accessors | 159 |
| Data transformation with the apply function | 134 | Calculating statistics | 166 |
| Filtering and querying | 137 | Calculating statistics by axis | 168 |
| Random sampling | 139 | Summary | 171 |

# 5

# Data Visualization with Plotly.js

| | | | |
|---|---|---|---|
| Technical requirements | 174 | Plotly.js | 196 |
| A brief primer on Plotly.js | 174 | Creating statistical charts with Plotly.js | 207 |
| Using Plotly.js via a script tag | 175 | Creating histogram plots with Plotly.js | 207 |
| Fundamentals of Plotly.js | 178 | Creating box plots with Plotly.js | 211 |
| Data format | 178 | Creating violin plots with Plotly.js | 215 |
| Configuration options for plots | 182 | Summary | 220 |
| Plotly layout | 190 | | |
| Creating basic charts with | | | |

# 6

# Data Visualization with Danfo.js

| | | | |
|---|---|---|---|
| Technical requirements | 222 | Creating scatter plots with Danfo.js | 229 |
| Setting up Danfo.js for plotting | 222 | Creating box and violin plots with Danfo.js | 231 |
| Adding Danfo.js to your code | 223 | Making box and violin plots for a Series | 232 |
| Downloading a dataset for plotting | 224 | Box and violin plots for multiple columns | 233 |
| Creating line charts with Danfo.js | 225 | | |

Box and violin plots with
specific x and y values                235

Creating histograms
with Danfo.js                          237
Creating a histogram from a Series     237
Creating a histogram from multiple
columns                                238

Creating bar charts
with Danfo.js                          240
Creating a bar chart from a Series     241
Creating a bar chart from multiple
columns                                242

Summary                                243

# 7

# Data Aggregation and Group Operations

Technical requirements                 247
Grouping data                          247
Single-column grouping                 247
Double-column grouping                 250

Iterating through grouped data 252
Iterating through single- and double-
column grouped data                    252

Using the .apply method                258

Data aggregation
of grouped data                        263
Data aggregation on
single-column grouping                 263
Data aggregation on
double-column grouping                 264
A simple application
of groupby on real data                265

Summary                                267

# Section 3: Building Data-Driven Applications

# 8

# Creating a No-Code Data Analysis/Handling System

Technical requirements                 272
Setting up the
project environment                    272
Structuring and designing
the app                                278
App layout and the
DataTable component                    282
Implementing DataTable components      284
File upload and state management       290

Creating different DataFrame
operation components                   294
Implementing the Describe component 295
Implementing the Query component    301
Implementing the Df2df component    306
Implementing the Arithmetic
component                              313

Implementing the
chart component                        317

Implementing the
ChartPlane component  318
Implementing the ChartViz component  321

Integrating ChartViz and ChartPlane
into App.js  322

**Summary**  **325**

# 9

# Basics of Machine Learning

**Technical requirements**  **328**
**Introduction to machine
learning**  **328**
A simple analogy of a machine learning
system  328

**Why machine learning works**  **331**
Objective functions  332
Evaluation metrics  333

**Machine learning problems/
tasks**  **334**

Supervised learning  335
Unsupervised learning  336

**Machine learning in JavaScript**  **337**
**Applications of machine
learning**  **338**
**Resources to understand
machine
learning in depth**  **339**
**Summary**  **339**

# 10

# Introduction to TensorFlow.js

**Technical requirements**  **342**
**What is TensorFlow.js?**  **342**
**Installing and using
TensorFlow.js**  **343**
Setting up TensorFlow.js in the browser  343
Installing TensorFlow.js in Node.js  345

**Tensors and basic operations
on tensors**  **348**
Creating tensors  350
Operating on tensors  352

**Building a simple regression
model with TensorFlow.js**  **357**
Setting up your environment locally  358
Retrieving and processing the
training dataset  359
Creating models with TensorFlow.js  363
Creating a simple three-layer
regression model  365
Training the model with the
processed dataset  366
Making predictions with
the trained model  369

**Summary**  **371**

# 11

# Building a Recommendation System with Danfo.js and TensorFlow.js

| | | | |
|---|---|---|---|
| Technical requirements | 374 | Building a movie recommendation system | 380 |
| What is a recommendation system? | 374 | Setting up your project directory | 381 |
| Collaborative filtering approach | 375 | Retrieving and processing the training dataset | 383 |
| Hybrid filtering approach | 376 | Building the recommendation model | 384 |
| The neural network approach to creating a recommendation system | 377 | Training and saving the recommendation model | 386 |
| | | Making movie recommendations with the saved model | 388 |
| | | Summary | 391 |

# 12

# Building a Twitter Analysis Dashboard

| | | | |
|---|---|---|---|
| Technical requirements | 394 | Creating the Search component | 404 |
| Setting up the project environment | 394 | Creating the ValueCounts component | 408 |
| Building the backend | 396 | Creating a plot component for sentiment analysis | 412 |
| Building the Twitter API | 396 | Creating a Table component | 414 |
| Building the text sentiment API | 400 | Summary | 419 |
| Building the frontend | 404 | | |

# 13

# Appendix: Essential JavaScript Concepts

| | | | |
|---|---|---|---|
| Technical requirements | 422 | Data types | 425 |
| Quick overview of JavaScript | 423 | Conditional branching and loops | 428 |
| Understanding the fundamentals of JavaScript | 424 | JavaScript functions | 435 |
| Declaring variables | 424 | Summary | 439 |

# Other Books You May Enjoy

# Index

# Preface

Most data analysts use Python and pandas for data processing and manipulation thanks to the convenience and performance that these libraries provide. However, JavaScript developers have always wanted **machine learning** (ML) to be possible in the browser as well. This book focuses on how Danfo.js brings data processing, analysis, and ML tools to JavaScript developers and how you can make the most of this library to develop data-driven applications.

The book starts with an introduction to JavaScript concepts and modern JavaScript. You'll then cover data analysis and transformation with Danfo.js and Dnotebook, which is an interactive computing environment for JavaScript. After that, the book covers how to load different types of datasets and analyze them by performing operations such as handling missing values, combining datasets, and string manipulations. You'll also focus on data plotting, visualization, data aggregation, and group operations by combining Danfo.js with Plotly. Later, you'll create a no-code data analysis and handling system with the help of Danfo.js. Then, you'll be introduced to basic ML concepts and how to use Tensorflow.js and Danfo.js to build a recommendation system. Finally, you'll build a Twitter analytics dashboard powered by Danfo.js.

By the end of this book, you'll be able to build and embed data analytics, visualization, and ML capabilities into any JavaScript app in server-side Node.js or the browser.

## Who this book is for

This book is for data science beginners, data analysts, and JavaScript developers who want to explore each stage of data analysis and scientific computing using a wide range of datasets. You'll also find this book useful if you are a data analyst, data scientist, or JavaScript developer looking to implement Danfo.js in your ML workflow. A working understanding of the JavaScript programming language, data science, and ML will assist you in understanding the key concepts covered in this book; however, an introductory JavaScript section has been provided in the first chapter and the appendix of the book.

# What this book covers

*Chapter 1, An Overview of Modern JavaScript*, discusses ECMA 6 syntax and the use of `import` statements, class methods, `extend` methods, and constructors. It also goes deep into an explanation of the `Promise` method, the use of `async` and `await` functions, and the `fetch` method. It also introduces how to set up an environment that supports modern JavaScript syntax, and proper versioning, as well as covering how to write unit tests.

*Chapter 2, Dnotebook - An Interactive Computing Environment for JavaScript*, dives into Dnotebook. For readers coming from the Python ecosystem, this is similar to Jupyter Notebook. We discuss how to use Dnotebook, how to create and delete cells, how to write Markdown in it, and how to save and share your notebook.

*Chapter 3, Getting Started with Danfo.js*, introduces Danfo.js and how to create DataFrames and series. It also introduces some essential methods for data analysis and processing.

*Chapter 4, Data Analysis, Wrangling, and Transformation*, looks at the practical use of Danfo.js for real-world datasets. Here, you'll learn how to load different types of datasets and analyze them by performing operations such as missing-value handling, calculating descriptive statistics, performing mathematical operations, combining datasets, and string manipulations.

*Chapter 5, Data Visualization with Plotly.js*, introduces data plotting and visualization. Here, you'll learn the basics of data visualization and plotting, and how to use Plotly.js for basic plotting.

*Chapter 6, Data Visualization with Danfo.js*, introduces data plotting and visualization with Danfo.js. Here, you'll learn how to use Danfo.js for creating plots directly on DataFrames or series. You will also learn how to customize Danfo.js plots.

*Chapter 7, Data Aggregation and Group Operations*, introduces group-by operations and how to perform them with Danfo.js, including how to group by one or more columns, how to use the provided group-by-aggregate functions, and how to create custom aggregate functions with `.apply`. We also show how the internals of the group-by operation work.

*Chapter 8, Creating a No-Code Data Analysis/Handling System*, shows what Danfo.js can enable us to do. In this chapter, we will create a no-code data-handling and analysis environment where the user can upload their data and then analyze and process their data artistically.

*Chapter 9, Basics of Machine Learning*, introduces ML in simple terms. It also shows you how ML can be done in the browser with the help of some ML JavaScript tools.

*Chapter 10, Introduction to TensorFlow.js*, introduces TensorFlow.js. It also shows how to perform basic mathematical operations and how to create, train, save, and reload an ML model. This chapter also shows how to integrate Danfo.js and Tensorflow.js to train a model effectively.

*Chapter 11, Building a Recommendation System with Danfo.js and TensorFlow.js*, shows you how to build a movie recommendation system using TensorFlow.js and Danfo.js. It shows you how to train models in Node.js and how to integrate them with the client side. It also shows how Danfo.js makes data preprocessing easy.

*Chapter 12, Building a Twitter Analysis Dashboard*, is where you will build a Twitter analytics dashboard using Danfo.js at the frontend and backend; the aim is to show how easy it is to use the same library throughout for your data analytics app, compared to using Python for the backend and JavaScript for the frontend, for instance.

*Chapter 13, Appendix: Essential JavaScript Concepts*, introduces the JavaScript programming language. Here, we introduce beginners to variable definitions, function creation, and the different ways to perform computations in JavaScript.

# To get the most out of this book

In this book, you will require basic knowledge of JavaScript and knowledge of frameworks such as Next.js, React.js, TensorFlow.js, and tailwindcss would be a plus.

| Software/hardware covered in the book | Operating system requirements |
| --- | --- |
| Dnotebook | Windows, macOS, or Linux |
| Chrome | Windows, macOS, or Linux |

**If you are using the digital version of this book, we advise you to type the code yourself or access the code from the book's GitHub repository (a link is available in the next section). Doing so will help you avoid any potential errors related to the copying and pasting of code.**

# Download the example code files

You can download the example code files for this book from GitHub at `https://github.com/PacktPublishing/Building-Data-Driven-Applications-with-Danfo.js`. If there's an update to the code, it will be updated in the GitHub repository.

We also have other code bundles from our rich catalog of books and videos available at `https://github.com/PacktPublishing/`. Check them out!

# Download the color images

We also provide a PDF file that has color images of the screenshots and diagrams used in this book. You can download it here: `https://static.packt-cdn.com/downloads/9781801070850_ColorImages.pdf`.

# Conventions used

There are a number of text conventions used throughout this book.

`Code in text`: Indicates code words in text, database table names, folder names, filenames, file extensions, pathnames, dummy URLs, user input, and Twitter handles. Here is an example: "In the case of the `financial_df` DataFrame, the index was auto-generated when we downloaded the dataset with the `read_csv` function."

A block of code is set as follows:

```
const df = new DataFrame({...})
df.plot("my_div_id").<chart type>
```

When we wish to draw your attention to a particular part of a code block, the relevant lines or items are set in bold:

```
...
var config = {
        displayModeBar: true,
        modeBarButtonsToAdd: [
...
```

Any command-line input or output is written as follows:

```
npm install @tensorflow/tfjs
```

**Bold**: Indicates a new term, an important word, or words that you see onscreen. For instance, words in menus or dialog boxes appear in **bold**. Here is an example: "In Microsoft Edge, open the Edge menu in the upper right-hand corner of the browser window and select **F12 Developer Tools**."

> **Tips or important notes**
> Appear like this.

# Get in touch

Feedback from our readers is always welcome.

**General feedback**: If you have questions about any aspect of this book, email us at customercare@packtpub.com and mention the book title in the subject of your message.

**Errata**: Although we have taken every care to ensure the accuracy of our content, mistakes do happen. If you have found a mistake in this book, we would be grateful if you would report this to us. Please visit www.packtpub.com/support/errata and fill in the form.

**Piracy**: If you come across any illegal copies of our works in any form on the internet, we would be grateful if you would provide us with the location address or website name. Please contact us at copyright@packt.com with a link to the material.

**If you are interested in becoming an author**: If there is a topic that you have expertise in and you are interested in either writing or contributing to a book, please visit authors.packtpub.com.

# Share Your Thoughts

Once you've read *Building Data-Driven Applications with Danfo.js*, we'd love to hear your thoughts! Scan the QR code below to go straight to the Amazon review page for this book and share your feedback.

https://packt.link/r/1-801-07085-7

Your review is important to us and the tech community and will help us make sure we're delivering excellent quality content.

# Section 1:
# The Basics

This section introduces JavaScript and the Node.js framework. These concepts are needed in order to fully understand and use Danfo.js. It also introduces how to set up an environment for modern JavaScript using Babel and Node.js, and also teaches the reader some basics of code testing.

This section comprises the following chapters:

- *Chapter 1, An Overview of Modern JavaScript*

# 1
# An Overview of Modern JavaScript

In this chapter, we will talk about some core JavaScript concepts. If you are new to JavaScript and need an introduction, check out *Chapter 13, Appendix: Essential JavaScript Concepts.*

Understanding some of the modern concepts of JavaScript is not a prerequisite to using Danfo.js, but we recommend going through this chapter if you're new to JavaScript or coming from a Python background, the reason being that we'll be using most of the concepts introduced here when building applications with Danfo.js. Also, it is worth mentioning that many of the concepts introduced here will, in general, help you to write better JavaScript.

This chapter introduces you to some modern JavaScript concepts, and by the end, you will have learned and understand the following concepts:

- Understanding the difference between let and var
- Destructuring
- Spread syntax
- Overview of scopes and closures
- Understanding Array and Object methods

- Understanding the this property

- Arrow functions

- Promises and async/await

- Object-oriented programming and JavaScript classes

- Setting up a modern JavaScript environment with transpilers

- Unit testing with Mocha and Chai

# Technical requirements

The major requirement is to have Node.js and NPM installed. You can follow the official installation guide available at `https://nodejs.org/en/download/` to install Node for your operating system. The code for this chapter can be found in the GitHub repo here: `https://github.com/PacktPublishing/Building-Data-Driven-Applications-with-Danfo.js/tree/main/Chapter01`.

We will start this section by understanding the difference between `let` and `var`, and why you should use `let` more often.

# Understanding the difference between let and var

Before ECMA 6, the common way of creating a variable was with the use of `var`. However, using `var` sometimes introduces bugs that mostly show up at runtime and others that are not revealed at runtime but may affect the way your code works.

Some of the properties of `var` that introduce bugs as mentioned in the previous paragraph are as follows:

- `var` allows the redeclaration of variables.

- `var` is not blocked scope; hence, it is either attached to the global scope or to a function scope.

Let's discuss the two properties listed above in detail.

## var allows the redeclaration of variables

var gives users access to redeclare variables along the line, hence overriding the previous variable of the same name. This feature might not show an error if not caught, but will certainly affect the behavior of the code:

```
var population_count = 490;
var new_count = 10;

//along the line; you mistakenly re-declare the variable
var population_count = "490"

//do some arithmetic operation with the variable
var total_count = population_count + new_count

//output: "49010"
```

In the preceding code snippet, there won't be any error, but the main objective of the code is altered just because var did not alert us that such a variable has been declared already.

Let's say we replace var with let, as shown in the following code:

```
let population_count = 490;
// ...some other code goes here
let population_count = "490"

//output: Error: Identifier population count as already being
declared
```

You can see from the preceding error output that let, unlike var, will not allow you to declare a variable in the same namespace twice.

Next, let's look at the scope property of variables declared with var.

## var is not a blocked scope

Variables declared with var have the following properties:

- They are readily available in the scope to which they are defined.
- They are available to scope within the range they are being declared.

In the following code, we will check how the `estimate` variable declared with `var` is accessible across all the scope within the variable declaration scope:

```
var estimate = 6000;
function calculate_estimate() {
    console.log(estimate);
}
calculate_estimate() // output 6000

if(true){
    console.log(estimate);
}
```

Now, for a blocked scope such as `if`, `while` loop, and `for` loop, the code within the blocked scope is meant to be run when the scope is available. Likewise, the variable is meant to exist only when the scope is available, and once the scope is not available again, the variable should not be accessible.

Declaring variables with `var` makes the preceding statement not possible. In the following code, we declare a variable using `var` and investigate its availability across all possible scopes:

```
if(true){
  var estimate = 6000;
}
console.log(estimate)
```

This will output the estimate as `6000`. The variable is not meant to exist outside the `if` block. Using `let` helps to solve this:

```
if(true){
  let estimate = 6000;
}
console.log(estimate)
//output: ReferenceError: estimate is not defined
```

This shows that using `let` to declare variables helps reduce unprecedented bugs in your code. In the next section, we'll discuss another important concept called destructuring.

# Destructuring

**Destructuring** is an assignment syntax in JavaScript that makes it easy to unpack values from arrays, or properties from objects, into distinct variables. For example, in the following code snippet, we can easily unpack the values 20, John, Doe, and 2019 into specified variables:

```
let data2 = [20, "John", "Doe", "2019"];
let [ age1, firstName1, lastName1, year1] = data2
```

Destructuring makes it possible to assign the element of an array to a variable, unlike the old conventional method of accessing an array element as shown in the following code:

```
//Old method of accessing an array
let data = [20, "John", "Doe", "2019"];

let firstName = data[1];
let age = data[0];
let lastName = data[2];
let year = data[3];
```

Destructuring also works on objects, as shown in the following code:

```
let data3 = {
    age: 20,
    firstName: "john",
    lastName: "Doe",
    year: 2019
}
let { age2, firstName2, lastName2, year2 } = data3
```

In object destructuring, notice that we use { } instead of [ ], as used for arrays. This is because the type on the left-hand side must be the same as the type on the right-hand side.

> **Important note**
>
> If we are to use [ ] while destructuring for an object, we receive an error, showing **TypeError**, while as a result of using { } for array destructuring, you might not obtain any error, but the variables will be undefined.

In the next section, we take a look at spread syntax.

# Spread syntax

**Spread syntax** is another form of destructuring for iterable elements such as strings and arrays. The spread syntax can be used in many situations involving arrays and objects. In this section, we'll quickly look at some of the use cases of spread syntax.

## Spreading or unpacking an iterable into an array

An iterable can be expanded/unpacked into an array. In the following example, we will show how to use the spread operator to unpack a string variable:

```
let name = "stephen"
let name_array = [...name];
```

The code expands the `name` string into `name_array`, hence, `name_array` will have the following values: `['s', 't', 'e','p', 'h', 'e','n']`.

While expanding the string element into the array, we can add other values alongside, as shown in the following code:

```
let name = "stephen"
let name_array = [...name, 1,2,3]
console.log(name_array)
// output ['s', 't', 'e','p', 'h', 'e','n',1,2,3]
```

Remember that any iterable can be spread into an array. This shows that we can also spread one array into another, as demonstrated in the following code:

```
let series = [1,2,3,4,5,6,7,8]
let new_array = [...series, 100, 200]
console.log(new_array)
// output [1, 2, 3, 4, 5,6, 7, 8, 100, 200]
```

Next, we'll apply the spread operator to objects.

# Creating new objects from existing ones

Creating new objects from existing ones follows the same pattern as the **Spread** operator:

```
Let data = {
    age: 20,
    firstName: "john",
    lastName: "Doe",
    year:  2019
}
let  new_data = {...data}
```

This creates a new object having the same property as the former object. While expanding the former object into the new one, new properties can be added alongside:

```
let data = {
    age: 20,
    firstName: "john",
    lastName: "Doe",
    year: 2019
}

let new_data = { ...data, degree: "Bsc", level: "expert" }
console.log(new_data)
//output
// {
//      age: 20,
//      Degree: "Bsc",
//      FirstName: "John",
//      lastName: "Doe",
//      Level: "expert",
//      Year: 2019
// }
```

## Function arguments

For functions requiring a lot of arguments, the spread syntax can help pass in a lot of arguments at once into the function, thereby reducing the stress of filling in the function's arguments one after the other.

In the following code, we will see how an array of arguments can be passed into a function:

```
function data_func(age, firstName, lastName, year) {
    console.log(`Age: ${age}, FirstName: ${firstName},
LastName: ${lastName}, Year: ${year}`);
}
let data = [30, "John", "Neumann", '1948']
data_func(...data)
//output Age: 30, FirstName: John, LastName: Neumann, Year:
1984
Age: 30, FirstName: John, LastName: Neumann, Year: 1984
```

In the preceding code, first, we created a function called `data_func` and defined a set of arguments to be passed in. We then created an array containing a list of parameters to pass to `data_func`.

By using spread syntax, we were able to pass the data array and assign each of the values in the array as an argument value – `data_func(...data)`. This becomes handy whenever a function takes many arguments.

In the next section, we will look at scope and closures, and how to use them to understand your JavaScript code better.

# Overview of scopes and closures

In the *Understanding the difference between let and var* section, we discussed scope and talked about how `var` is available in the global scope, as well as in the function scope. In this section, we will be moving into scope and closures in a little more depth.

## Scope

To understand scope, let's start with the following code:

```
let food = "sandwich"
function data() {

}
```

The food variables and data function are both assigned to the global scope; hence, they are termed **a global variable** and **global function**. These global variables and functions are always accessible to every other scope and program in the JavaScript file.

The **local** scope can further be grouped as follows:

- **Function scope**
- **Block scope**

Function scope is only available within a function. That is, all variables and functions created within a function scope are not accessible outside the function, and only exist when the function scope is available, for example:

```
function func_scope(){
// function scope exist here
}
```

The block scope exists in specific contexts only. For instance, it can exist within a curly brace, { }, along with the if statement, for loop, and while loop. Two more examples are presented in the following code snippets:

```
if(true){
// if block scope
}
```

In the preceding if statement, you can see that the block scope only exists inside the curly braces, and all variables declared inside the if statement are local to it. Another example is a for loop, as shown in the following code snippet:

```
for(let i=0; i< 5; i++){
//for loop's block scope
}
```

The block scope also exists inside the curly braces of a `for...` loop. Here, you have access to the `i` counter, and any variables declared inside cannot be accessed outside the block.

Next, let's understand the concept of closures.

## Closure

**Closure** makes use of the idea of scope within functions. Remember we agreed that the variables declared within a function scope are not accessible outside the function scope. Closure gives us the ability to make use of these private properties (or variables).

Let's say we want to create a program that will always add the values 2 and 1 to an `estimate` variable representing a population estimate. One way to do this is shown in the following code:

```
let estimate = 6000;
function add_1() {
    return estimate + 1
}
function add_2() {
    return estimate + 2;
}
console.log(add_1()) // 60001
console.log(add_2()) // 60002
```

There's nothing wrong with the preceding code, but as the code base becomes very big, we might lose track of the `estimate` value, perhaps a function along the line to update the value, and we may also want to make the global scope clean by making the global `estimate` variable a local variable.

Hence, we can create a function scope to do this for us and ultimately, clean the global scope. Here is an example in the following code snippet:

```
function calc_estimate(value) {
    let estimate = value;
    function add_2() {
        console.log('add two', estimate + 2);
    }
    function add_1() {
        console.log('add one', estimate + 1)
    }
```

```
  add_2();
  add_1();


}
calc_estimate(6000) //output: add two 60002 , add one 60001
```

The preceding code snippet is similar to the first one we defined, with just a tiny difference, that is, the function accepts the estimate value and then creates the add_2 and add_1 functions inside the calc_estimate function.

A better way to showcase closure using the preceding code is to have the ability to update the estimate value whenever we want and not at the instance where the function is called. Let's see an example of this:

```
function calc_estimate(value) {
  let estimate = value;
  function add_2() {
    estimate += 2
    console.log('add 2 to estimate', estimate);
  }
  return add_2;
}
let add_2 = calc_estimate(50);
// we have the choice to add two to the value at any time in
our code
add_2() // add 2 to estimate 52
add_2() // add 2 to estimate 54
add_2() // add 2 to estimate 56
```

In the preceding code snippet, the inner function, add_2, will add the value 2 to the estimate variable, thereby changing the value. calc_estimate is called and assigned to a variable, add_2. With this, whenever we call add_2, we update the estimated value by 2.

We update the add_2 function inside calc_estimate to accept a value that can be used to update the estimate value:

```
function calc_estimate(value){
  let estimate = value;
  function add_2(value2){
```

```
    estimate +=value2
    console.log('add 2 to estimate', estimate);
  }
  return add_2;
}
let add_2 = calc_estimate(50);
// we have the choice to add two to the value at any time in
our code

add_2(2) // add 2 to estimate 52
add_2(4) // add 2 to estimate 56
add_2(1) // add 2 to estimate 5
```

Now that you've learned about scopes and closures, we will move to arrays, objects, and string methods in the following section.

> **Further reading**
> To go into closures in greater detail, check out the book *Mastering JavaScript*, by *Ved Antani*.

# Understanding Array and Object methods

**Arrays** and **Objects** are the two most important data types in JavaScript. As such, we have dedicated a section to talking about some of their methods. We will start with the Array methods.

## Array methods

We can't discuss how to build a data-driven product without discussing Array methods. Knowing different Array methods gives us the privilege of accessing our data and creating tools to manipulate/handle our data.

An array can be created in two different forms:

```
let data = []
// or
let data = new Array()
```

The [  ] method is mostly used for initializing arrays, while the new `Array()` method is used mostly to create an empty array of an *n* size, as shown in the following code snippet:

```
let data = new Array(5)
console.log(data.length) // 5
console.log(data) //  [empty x 5]
```

The empty array created can later be filled with values as shown in the following code:

```
data[0] = "20"
data[1] = "John"
data[2] = "Doe"
data[3] = "1948"
console.log(data) // ["20", "John","Doe","1948", empty]
// try access index 4
console.log(data[4]) //  undefined
```

Creating such an empty array is not limited to using the new `Array()` method. It can also be created with the [  ] method, as shown in the following code snippet:

```
let data = []
data.length = 5; // create an empty array of size 5
console.log(data)  // [empty x 5]
```

You can see that we explicitly set the length after creation, and as such, the new `Array()` method is more convenient.

Let's now look at some of the common array methods that will be used in building some of our data-driven tools.

## Array.splice

Deleting and updating array values will always be one of the essential things in a data-driven product. JavaScript has a `delete` keyword to delete a value at a particular index in an array. This method does not actually delete the value, but replaces it with an empty or undefined value, as shown in the following code:

```
let data = [1,2,3,4,5,6];
delete data[4];
console.log(data) // [1,2,3,4 empty, 6]
```

In the `data` variable, if we try to access the value at index 4, we will see that it returns `undefined`:

```
console.log(data[4]) // undefined
```

But whenever we use `splice` to delete a value in an array, the index of the array is re-arranged, as demonstrated in the following code snippet:

```
let data = [1,2,3,4,5,6]
data.splice(4,1) // delete index 4
console.log(data) // [1,2,3,4,6]
```

`Array.splice` takes in the following argument, `start`, `[deleteCount, value-1,......N-values]`. In the preceding code snippet, since we are only deleting, we make use of `start` and `deleteCount`.

The `data.splice(4,1)` command deletes the value starting at index 4, with only one count, hence it deletes the value at index 5.

If we replace the value `1` in `data.splice(4,1)` with `2`, resulting in `data.splice(4,2)`, two values (5 and 6) from the `data` array will be deleted, starting from index 4, as shown in the following code block:

```
let data = [1,2,3,4,5,6]
data.splice(4,0,10,20) // add values between 5 and 6
console.log(data) // [1,2,3,4,5,10,20,6]
```

`data.splice(4,0,10, 20);` specifies starting at index 4, and `0` specifies that no values should be deleted, while adding the new values (`10` and `20`) between 5 and 6.

## Array.includes

This method is used for checking whether an array contains a particular value. We show an example in the following code snippet:

```
let data = [1,2,3,4,5,6]
data.includes(6) // true
```

## Array.slice

`Array.slice` is used to obtain an array element by specifying the range; `Array.slice(start-index, end-index)`. Let's see an example of using this method in the following code:

```
let data = [1,2,3,4,5,6]
data.slice(2,4)
//output [3,4]
```

The preceding code extracts elements, starting from index 2 (having element 3) to index 5. Note that the array did not output `[3,4,5]`, but `[3,4]`. `Array.splice` always excludes the end index value, and so it uses a close end range.

## Array.map

The `Array.map` method iterates through all the elements of an array, applies some operations to each iteration, and then returns the result as an array. The following code snippet is an example:

```
let data = [1,2,3,4,5,6]
let data2 = data.map((value, index)=>{
return value + index;
});
console.log(data2) // [1,3,5,7,9,11]
```

The `data2` variable is created by iterating over each data element by using a map method. In the `map` method, we are adding each element (value) of the array to its index.

## Array.filter

The `Array.filter` method is used to filter out some elements in an array. Let's see this in action:

```
let data = [1,2,3,4,5,6]
let data2 = data.filter((elem, index)=>{
return (index %2 == 0)
})
console.log(data2) // [1,3,5]
```

In the preceding code snippet, the array element of data at every even index is filtered out using the modulus (%) of 2.

There are lots of Array methods, but we covered these few methods because they are always handy during data handling, and we will not be able to cover all of them.

However, if any new method is used in later chapters of this book, we will certainly provide an explanation. In the next section, we'll discuss Object methods.

# Objects

**Objects** are the most powerful and important data type in JavaScript, and in this section, we'll introduce some important properties and methods of objects that make working with them easier.

## Accessing object elements

Accessing keys/values in an object is important, so there exists a special for . . . in loop for doing that:

```
for (key in object) {
    // run some action with keys
}
```

The for . . . in loop returns all the keys in an object, and this can be used to access Object values, as demonstrated in the following code:

```
let user_profile = {
  name: 'Mary',
  sex: 'Female',
  age: 25,
  img_link: 'https://some-image-link.png',
}
for (key in user_profile) {
    console.log(key, user_profile[key]);
}
//output:
// name Mary
// sex Female
// age 25
// img_link https://some-image-link.png
```

In the next section, we will show how to test the existence of properties.

## Testing for the existence of property

To check whether a property exists, you can use the `"key"` in object syntax, as demonstrated in the following code snippet:

```
let user_profile = {
  name: 'Mary',
  sex: 'Female',
  age: 25,
  img_link: 'https://some-image-link.png',
}

console.log("age" in user_profile)
//outputs: true

if ("rank" in user_profile) {
    console.log("Your rank is", user_profile.rank)
} else {
    console.log("rank is not a key")
}
//outputs: rank is not a key
```

## Deleting properties

The `delete` keyword used before an object property will remove a specified property from an object. Look at the following example:

```
let user_profile = {
    name: 'Mary',
    sex: 'Female',
    age: 25,
    img_link: 'https://some-image-link.png',
}
delete user_profile.age
console.log(user_profile)
//output:
// {
//     img_link: "https://some-image-link.png",
```

```
//      name: "Mary",
//      sex: "Female"
// }
```

You can see that the age property has been successfully removed from the user_ profile object. Next, let's look at how to copy and clone objects.

## Copying and cloning objects

Assigning an old object to a new one simply creates a reference to the old object. That is, any modification made to the new object also affects the old one. For instance, in the following example, we assign the user_profile object to a new variable, new_user_ profile, and then proceed to delete the age property:

```
let user_profile = {
    name: 'Mary',
    sex: 'Female',
    age: 25,
    img_link: 'https://some-image-link.png',
}
let new_user_profile = user_profile
delete new_user_profile.age

console.log("new_user_profile", new_user_profile)
console.log("user_profile", user_profile)
//output:
// "new_user_profile" Object {
//      img_link: "https://some-image-link.png",
//      name: "Mary",
//      sex: "Female"
// }

// "user_profile" Object {
//      img_link: "https://some-image-link.png",
//      name: "Mary",
//      sex: "Female"
// }
```

You will notice that deleting the `age` property from the `user_profile` object also deletes it from `new_user_profile`. This is because the copy is simply a reference to the old object.

In order to copy/clone objects as new and independent ones, you can use the `Object.assign` method, as shown in the following code:

```
let new_user_profile = {}
Object.assign(new_user_profile, user_profile)

delete new_user_profile.age

console.log("new_user_profile", new_user_profile)
console.log("user_profile", user_profile)

//output
"new_user_profile" Object {
  img_link: "https://some-image-lik.png",
  name: "Mary",
  sex: "Female"
}
"user_profile" Object {
  age: 25,
  img_link: "https://some-image-lik.png",
  name: "Mary",
  sex: "Female"
}
```

The `Object.assign` method can also be used to copy properties from more than one object at the same time. We present an example in the following code snippet:

```
let user_profile = {
  name: 'Mary',
  sex: 'Female',
  age: 25,
  img_link: 'https://some-image-lik.png',
}
let education = { graduated: true, degree: 'BSc' }
```

```
let permissions = { isAdmin: true }

Object.assign(user_profile, education, permissions);
console.log(user_profile)
//output:
// {
//      name: 'Mary',
//      sex: 'Female',
//      img_link: 'https://some-image-link.png',
//      graduated: true,
//      degree: 'BSc',
//      isAdmin: true
// }
```

You can see that we were able to copy properties from two objects (education and permissions) into our original object, user_profile. In this way, we can copy any number of objects into another one by simply listing all the objects when you call the Object.assign method.

> **Tip**
>
> You can also perform a deep copy with a **spread** operator. This is actually quicker and easier to write, as demonstrated in the following example:
>
> ```
> let user_profile = {
>   name: 'Mary',
>   sex: 'Female'
> }
>
> let education = { graduated: true, degree: 'BSc' }
> let permissions = { isAdmin: true }
> const allObjects = {...user_profile, ...education,
> ...permissions}
> ```
>
> Spreading each object, as shown in the preceding code snippet, performs a deep copy of all properties in allObjects. This syntax is easier and quicker than the object.assign method and is largely used today.

In the next section, we will talk about another important concept relating to JavaScript objects called the **this** property.

# Understanding the this property

The **this** keyword is an object property. When used within a function, it takes the form of the object to which the function is bound at invocation.

In every JavaScript environment, we have a global object. In Node.js, the global object is named **global** and, in the browser, the global object is named **window**.

By global object, we mean that all the variable declarations and functions are represented as a property and method of this global object. For example, in a browser script file, we can access the global objects, as shown in the following code snippet:

```
name = "Dale"
function print() {
    console.log("global")
}
// using the browser as our environment
console.log(window.name) // Dale
window.print() // global
```

In the preceding code block, the `name` variable and `print` function are declared at the global scope, hence they can be accessed as an attribute (`window.name`) and method (`window.print()`) of the **window** global object.

The statement made in the previous sentence can be summarized as the global name and function are binded (or assigned) by default to the global object window.

This also means that we can always bind this variable to any object having the same `name` variable and the same function, called `print`.

To get this concept, first, let's re-write `window.print()` as `print.call(window)`. This new method is called de-sugaring in JavaScript; it is like seeing an implementation of a method in its real form.

The `.call` method simply takes in the object we want to bind a function call to.

Let's see how `print.call()` and how this property works. We'll rewrite the `print` function to access the `name` variable, as shown in the following code snippet:

```
name  = "Dale"
object_name = "window"
function print(){
   console.log(`${this.name} is accessed from        ${this.
object_name}`)
}
console.log(print.call(window))  // Dale is accessed from window
```

Now, let's create a custom object and also give it the same property as the `window` object, as shown in the following code snippet:

```
let custom_object = {
name: Dale,
Object_name: "custom_object"
}


print.call(custom_object) // Dale is accessed from custom_
object
```

This concept can be applied to all Object methods, as shown in the following code:

```
data = {
             name: 'Dale',
             obj_name: 'data',
             print: function () {
                  console.log(`${this.name} is accessed from
${this.obj_name}`);
             }
      }
data.print() // Dale is accessed from data
// don't forget we can also call print like this
data.print.call(data) // Dale is accessed from data
```

With this, we can also bind the `print()` method from `data` to another object, as shown in the following code snippet:

```
let data2 = {
  name: "Dale D"
  Object_name: "data2"
}
data.print.call(data2) // Dale D is accessed from data2
```

This method shows how this property depends on the function invocation runtime. This concept also affects how some event operations work in JavaScript.

> **Further reading**
>
> To get a deeper understanding of this concept, Yehuda Katz, one of the creators of *Emberjs and Members of TC39*, sheds more light on this in his article, *Understanding JavaScript Function Invocation and "this"*.

# Arrow functions

Arrow functions are just unnamed or anonymous functions. The general syntax of **arrow** functions is shown in the following expression:

```
( args ) => { // function body }
```

Arrow functions provide a means of creating concise callable functions. By this, we mean arrow functions are not constructible, that is, they can't be instantiated with the new keyword.

The following are different ways of how and when to use arrow functions:

- The arrow function can be assigned to a variable:

```
const unnamed = (x) => {
  console.log(x)
}
unnamed(10) //  10
```

- Arrow functions can be used as an **IIFE (Immediately Invoked Function Expression)**. IIFEs are functions that once encountered by the JavaScript compiler are called immediately:

```
((x) => {
    console.log(x)
})("unnamed function as IIFE") // output: unnamed
function as IIFE
```

- Arrow functions can be used as callbacks:

```
function processed(arg, callback) {
    let x = arg * 2;
    return callback(x);
}
processed(2, (x) => {
    console.log(x + 2)
});    // output:   6
```

While arrow functions are great in some situations, there is a downside to using them. For example, arrow functions do not have their own `this` scope, hence its scope is always bound to the general scope, thereby changing our whole idea of function invocation.

In the *Understanding the this property* section, we talked about how functions are bounded to their invocation scope and using this ability to support **closure**, but using the arrow function denies us this feature by default:

```
const Obj = {
    name: "just an object",
    func: function(){
        console.log(this.name);
    }
}
Obj.func() // just an object
```

Even though in the object, as shown in the code snippet, we make use of the anonymous function (but not the arrow function), we have access to the object's `Obj` properties:

```
const Obj = {
    name: "just an object",
    func:  () => {
```

```
            console.log(this.name);
      }
}
Obj.func() // undefined
```

The arrow function used makes the `Obj.func` output `undefined`. Let's see how it works if we have a variable called name in the global scope:

```
let name = "in the global scope"
const Obj = {
      name: "just an object",
      func:   () => {
            console.log(this.name);
      }
}

Obj.func() // in the global
```

As we can see, `Obj.func` makes a call to the variable in the global scope. Hence, we must know when and where to use the arrow functions.

In the next section, we will talk about Promises and async/await concepts. This will give us the power to easily manage long-running tasks and avoid callback hell (callbacks having callbacks).

# Promises and async/await

Let's dive a bit into the world of Asynchronous functions, functions that we call now but finish later. In this section, we will see why we need **Promise** and **async/await**.

Let's start with a simple problem as shown in the following code snippet. We are given a task to update an array with a function, after 1 second of calling the function:

```
let syncarray = ["1", "2", "3", "4", "5"]
function addB() {
    setTimeout(() => {
        syncarray.forEach((value, index)=>{
            syncarray[index] = value + "+B"
        })
        console.log("done running")
```

```
    }, 1000)
}
addB()
console.log(syncarray);
// output
// ["1", "2", "3", "4", "5"]
// "done running"
```

`console.log(syncarray)` is executed before the `addB()` function, hence we see the `syncarray` output before it is being updated. This is an Asynchronous behavior. One of the ways to solve this is to use a callback:

```
let syncarray = ["1", "2", "3", "4", "5"]
function addB(callback) {
    setTimeout(() => {
        syncarray.forEach((value, index)=>{
            syncarray[index] = value + "+B"
        })
        callback() //call the callback function here
    }, 1000)
}
addB(()=>{
    // here we can do anything with the updated syncarray
    console.log(syncarray);
})
// output
// [ '1+B', '2+B', '2+B', '4+B', '5+B' ]
```

Using the preceding callback approach means that we always pass in callbacks in order to perform other operations on the updated `syncarray` function. Let's update the code a little, and this time we'll also add the string `"A"` to `syncarray` and then print out the updated array:

```
let syncarray = ["1", "2", "3", "4", "5"]
function addB(callback) {
    setTimeout(() => {
        syncarray.forEach((value, index) => {
```

```
                syncarray[index] = value + "+B"
            })
        callback() //call the callback function here
    }, 1000)
}
addB(() => {
    setTimeout(() => {
        syncarray.forEach((value, index) => {
            syncarray[index] = value + "+A";
        })
        console.log(syncarray);
    }, 1000)
})
// output
// [ '1+B+A', '2+B+A', '3+B+A', '4+B+A', '5+B+A' ]
```

The preceding code block shows a quick way of passing `callback`. Based on the arrow function we discussed, it can be more organized by creating a named function.

## Cleaning callbacks with promises

Using callbacks quickly becomes unwieldy and can quickly descend into callback hell. One method of freeing ourselves from this is to make use of Promises. Promises makes our callbacks more organized. It gives a chainable mechanism to unify and orchestrate code that is dependent on previous functions, as you'll see in the following code block:

```
let syncarray = ["1", "2", "3", "4", "5"]
function addA(callback) {
    return new Promise((resolve, reject) => {
        setTimeout(() => {
            syncarray.forEach((value, index) => {
                syncarray[index] = value + "+A";
            })
            resolve()
        }, 1000);
    })
}
```

```
addA().then(() => console.log(syncarray));
//output
//[ '1+A', '2+A', '2+A', '4+A', '5+A' ]
```

In the preceding code snippet, setTimeout is wrapped inside the Promise function. A Promise is always instantiated using the following expression:

```
New Promise((resolve, rejection) => {
})
```

A Promise is either resolved or rejected. When it is resolved, then we are free to do other things, and when it is rejected, we need to handle the error.

For example, let's ensure that the following Promise is rejected:

```
let syncarray = ["1", "2", "3", "4", "5"]
function addA(callback) {
    return new Promise((resolve, reject) => {
        setTimeout(() => {
            syncarray.forEach((value, index) => {
                syncarray[index] = value + "+A";
            })
            let error = true;
            if (error) {
                reject("just testing promise rejection")
            }
        }, 1000);
    })
}
addA().catch(e => console.log(e)) // just testing promise
rejection
```

And whenever we have multiple promises, we can use the .then() method to handle each one:

```
addA.then(doB)
    .then(doC)
    .then(doD)
    .then(doF)
    .catch(e= > console.log(e));
```

The use of multiple `.then()` methods to handle numerous promises can quickly become unwieldy. To prevent this, we can use methods such as `Promise.all()`, `Promise.any()`, and `Promise.race()`.

The `Promise.all()` method takes in an array of promises to be executed, and will only resolve when all promises are fulfilled. In the following code snippet, we add another Asynchronous function to our previous example and use `Promise.all()` to handle them:

```
let syncarray = ["1", "2", "2", "4", "5"]
function addA() {
    return new Promise((resolve, reject) => {
        setTimeout(() => {
            syncarray.forEach((value, index) => {
                syncarray[index] = value + "+A";
            })
            resolve()
        }, 1000);
    })
}
function addB() {
    return new Promise((resolve, reject) => {
        setTimeout(() => {
            syncarray.forEach((value, index) => {
                syncarray[index] = value + "+B";
            })
            resolve()
        }, 2000);
    })
}
Promise.all([addA(), addB()])
.then(() => console.log(syncarray)); // [ '1+A+B', '2+A+B',
'2+A+B', '4+A+B', '5+A+B' ]
```

From the output in the preceding section, you can see that each Asynchronous function gets executed in the order it was added, and the final result is the effect of both functions on the `syncarray` variable.

The `promise.race` method, on the other hand, will return as soon as any promise in the array is resolved or rejected. You can think of this as a race where each promise tries to resolve or reject first, and as soon as this happens, the race is over. To see an in-depth explanation as well as code examples, you can visit the MDN docs here: `https://developer.mozilla.org/en-US/docs/Web/JavaScript/Reference/Global_Objects/Promise/any`.

And finally, the `promise.any` method will return on the first fulfilled promise irrespective of any other rejected `promise` function. If all promises are rejected, then `Promise.any` rejects promises by providing errors for all of them. To see an in-depth explanation as well as code examples, you can visit the MDN docs here: `https://developer.mozilla.org/en-US/docs/Web/JavaScript/Reference/Global_Objects/Promise/race`.

While using promises to work with callback solves a lot of issues, there is an even better way of implementing or using them. These are called **async/await** functions. We'll introduce these functions and show you how to use them in the following section.

## async/await

As said earlier, async/await provides a more elegant way of working with promises. It gives us the power to control how and when each promise function gets called inside a function, instead of using `.then()` and `Promise.all()`.

The following code snippet shows how you can use async/await in your code:

```
Async function anyName() {
    await anyPromiseFunction()
        await anyPromiseFunction()
}
```

The preceding `async` function can contain as many promise functions as possible, each waiting for the other to execute before being executed. Also, note that an `async` function is resolved as a `Promise`. that is, you can only obtain the return variable of the preceding anyName function (or resolve the function) using `.then()` or by calling it in another async/await function:

```
Async function someFunction() {
    await anyPromiseFunction()
        await anotherPromiseFunction()
    return "done"
}
```

```
// To get the returned value, we can use .then()
anyName().then(value => console.log(value)) // "done"
// we can also call the function inside another Async/await
function
Async function resolveAnyName() {
   const result = await anyName()
   console.log(result)
}
resolveAnyName() // "done"
```

With this knowledge, here is how we can rewrite the promise execution from the previous section instead of using `Promise.all([addA(), addB()])`:

```
let syncarray = ["1", "2", "2", "4", "5"]
function addA(callback) {
    return new Promise((resolve, reject) => {
        setTimeout(() => {
            syncarray.forEach((value, index) => {
                syncarray[index] = value + "+A";
            })
            resolve()
        }, 1000);
    })
}
function addB(callback) {
    return new Promise((resolve, reject) => {
        setTimeout(() => {
            syncarray.forEach((value, index) => {
                syncarray[index] = value + "+B";
            })
            resolve()
        }, 2000);
    })
}
Async function runPromises(){
    await addA()
    await  addB()
```

```
        console.log(syncarray);
    }
  runPromises()
  //output: [ '1+A+B', '2+A+B', '2+A+B', '4+A+B', '5+A+B' ]
```

You can see from the preceding output that we have the same output as when using the `Promise.all` syntax, but are adopting a minimal and cleaner approach.

> **Note**
>
> One drawback of using multiple awaits as opposed to `promise.all` is efficiency. Though minor, `promise.all` is the preferred and recommended way to handle multiple independent promises.
>
> This thread (`https://stackoverflow.com/questions/45285129/any-difference-between-await-promise-all-and-multiple-await`) on Stack Overflow clearly explains why this is the recommended way to handle multiple promises.

In the next section, we'll discuss **object-oriented programming** (**OOP**) in JavaScript, and how to use ES6 classes.

# Object-oriented programming and JavaScript classes

OOP is a common programming paradigm supported by most high-level languages. In OOP, you typically program an application using the concept of objects, which can be a combination of data and code.

Data represents information about the object, while code represents attributes, properties, and behaviors that can be carried out on objects.

OOP opens up a whole new world of possibilities as many problems can be simulated or designed as the interaction between different objects, thereby making it easier to design complex programs, as well as maintain and scale them.

JavaScript, like other high-level languages, provides support for OOP concepts, although not fully (`https://developer.mozilla.org/en-US/docs/Web/JavaScript/Reference/Classes`), but in essence, most of the important concepts of OOP, such as **objects**, **classes**, and **inheritance**, are supported, and these are mostly enough to solve many problems you wish to model using OOP. In the following section, we will briefly look at classes and how these are related to OOP in JavaScript.

# Classes

Classes in OOP act like a blueprint for an object. That is, they define a template of an abstract object in such a way that multiple copies can be made by following that blueprint. Copies here are officially called **instances**. So, in essence, if we define a class, then we can easily create multiple instances of that class.

In ECMA 2015, the **class** keyword was introduced to JavaScript, and this greatly simplified the creation of classes in the language. The following code snippet shows how to model a User object using the ES16 class keyword:

```
class User {
    constructor(firstName, lastName, email) {
        this.firstName = firstName;
        this.lastName = lastName;
        this.email = email;
    }
    getFirstName() {
        return this.firstName;
    }
    getLastName() {
        return this.lastName;
    }
    getFullName() {
        return `${this.firstName} ${this.lastName}`;
    }
    getEmail() {
        return this.email;
    }
    setEmail(email) {
        this.email = email;
    }
}
let Person1 = new User("John", "Benjamin", "john@some-email.com")
console.log(Person1.getFullName());
console.log(Person1.getEmail());
// outputs
```

```
// "John Benjamin"
// "john@someemail.com"
```

By using the `class` keyword, you can wrap both data (names and email) with functionality (functions/methods) in a cleaner way that aids easy maintenance as well as understanding.

Before we move on, let's break down the class template in more detail for a better understanding.

The first line starts with the `class` keyword and is usually followed by a class name. A class name, by convention, is written in camel case, for instance, `UserModel` or `DatabaseModel`.

An optional constructor can be added inside a `class` definition. A `constructor` class is an initialization function that runs every time a new instance is created from a class. Here, you'll normally add code that initializes each instance with specific properties. For instance, in the following code snippet, we create two instances from the `User` class, and initialize them with specific properties:

```
let Person2 = new User("John", "Benjamin", "john@some-email.
com")
let Person3 = new User("Hannah", "Joe", "hannah@some-email.
com")
console.log(Person2.getFullName());
console.log(Person3.getFullName());
//outputs
// "John Benjamin"
// "Hannah Montanna"
```

The next important part of a class is the addition of functions. Functions act as `class` methods and generally add a specific behavior to the class. Functions are also available to every instance created from the class. In our `User` class, methods such as `getFirstName`, `getLastName`, `getEmail`, and `setEmail` are added to perform different functions based on their implementation. To call functions on class instances, you typically use a dot notation, as you would when accessing an object's property. For example, in the following code, we return the full name of the `Person1` instance:

```
Person1.getFullName()
```

With classes out of the way, we now move to the next concept in OOP, called *inheritance*.

# Inheritance

**Inheritance** in OOP is the ability of one class to use properties/methods of another class. It is an easy way of extending the characteristics of one class (subclass/child class) using another class (superclass/parent class). In that way, the child class inherits all the characteristics of the parent class and can either extend or change these properties. Let's use an example to better understand this concept.

In our application, let's assume we already have the `User` class defined in the previous section, but we want to create a new set of users called `Teachers`. Teachers are also a class of users, and they will also require basic properties, such as the name and email that the `User` class already has. So, instead of creating a new class with these existing properties and methods, we can simply extend it, as shown in the following code snippet:

```
class Teacher extends User {
}
```

Note that we use the `extends` keyword. This keyword simply makes all the properties in the parent class (`User`) available to the child class (`Teacher`). With just the basic setup, the `Teacher` class automatically has access to all the properties and methods of the `User` class. For instance, we can instantiate and create a new `Teacher` in the same way we created a `User` value:

```
let teacher1 = new Teacher("John", "Benjamin", "john@someemail.
com")
console.log(teacher1.getFullName());
//outputs
// "John Benjamin"
```

After extending a class, we basically want to add new features. We can do this by simply adding new functions or properties inside the child class template, as shown in the following code:

```
class Teacher extends User {
  getUserType() {
    return "Teacher"
  }
}
```

In the preceding code snippet, we added a new method, `getUserType`, which returns a string of the user type. In this way, we can add more features that were not originally in the parent class.

It is worth mentioning that you can replace parent functions in the `child` class by creating a new function in the `child` class with the same name. This process is called **method overriding**. For instance, to override the `getFullName` function in the `Teacher` class, we can do the following:

```
class User {
    constructor(firstName, lastName, email) {
        this.firstName = firstName;
        this.lastName = lastName;
        this.email = email;
    }
    getFirstName() {
        return this.firstName;
    }
    getLastName() {
        return this.lastName;
    }
    getFullName() {
        return `${this.firstName} ${this.lastName}`;
    }
    getEmail() {
        return this.email;
    }
    setEmail(email) {
        this.email = email;
    }
}
class Teacher extends User {
    getFullName(){
        return `Teacher: ${this.firstName} ${this.lastName}`;
    }
    getUserType(){
        return "Teacher"
    }
}
let teacher1 = new Teacher("John", "Benjamin", "john@someemail.
com")
```

```
console.log(teacher1.getFullName());
//output
// "Teacher: John Benjamin"
```

A question may arise here: What if we want to initialize the `Teacher` class with additional instances besides `firstname`, `lastname`, and `email`? This is achievable, and we can easily extend the constructor function by using a new keyword, `super`. We demonstrate how to do this in the following code:

```
// class User{
// previous User class goes here
//     ...
// }


class Teacher extends User {
    constructor(firstName, lastName, email, userType, subject)
{
        super(firstName, lastName, email) //calls parent class
constructor
        this.userType = userType
        this.subject = subject
    }
    getFullName() {
        return `Teacher: ${this.firstName} ${this.lastName}`;
    }
    getUserType() {
        return "Teacher"
    }
}
let teacher1 = new Teacher("Johnny", "Benjamin", "john@
someemail.com", "Teacher", "Mathematics")
console.log(teacher1.getFullName());
console.log(teacher1.userType);
console.log(teacher1.subject);
//outputs
// "Teacher: Johnny Benjamin"
// "Teacher"
// "Mathematics"
```

In the preceding code, we are performing two new things. First, we add two new instance properties (`userType` and `subject`) to the `Teacher` class, and then we are calling the `super` function. The `super` function simply calls the parent class (`User`), and performs the instantiation, and immediately after, we initialize the new properties of the `Teacher` class.

In this way, we are able to first initialize the parent properties before initializing the class properties.

Classes are very useful in OOP and the `class` keyword provided in JavaScript makes working with OOP easy. It is worth mentioning that under the hood, JavaScript converts the classes template to an object, as it does not have first-class support for classes. This is because JavaScript, by default, is a prototype-based, object-oriented language. Hence, the class interface provided is called **syntactic sugar** over the underlying prototype-based model, which JavaScript calls under the hood. You can read more about this at the following link: `http://es6-features.org/#ClassDefinition`.

Now that we have a basic understanding of OOP in JavaScript, we are ready to create complex applications that can be easily maintained. In the next section, we will discuss another important aspect of JavaScript development, which is setting up a development environment with modern JavaScript support.

# Setting up a modern JavaScript environment with transpilers

One of the unique features of JavaScript, and the reason why it is very popular, is its cross-platform support. JavaScript runs almost everywhere, from browsers and desktops to even on the server side. While this is a unique feature, getting JavaScript to run optimally in these environments requires some setup and configuration using third-party tools/libraries. Another reason why you need to set up tooling is that you can write JavaScript in different flavors, and because these modern/newer flavors may not be supported by older browsers. This means that the code you write in newer syntax, typically post-ES15, will need to be transpiled into pre-ES16 format for it to run properly in most browsers.

In this section, you will learn how to set up and configure a JavaScript project to support cross-platform and modern JavaScript code. You will use two popular tools – **Babel** and **webpack** – to achieve this.

# Babel

Babel is a tool for converting JavaScript code written in ES15 code into a backward-compatible version of JavaScript in modern or older browsers. Babel can help you to do the following:

- Transform/transpile syntax.
- Polyfill features that are missing in your target environment. Some modern features that are not available in older environments are automatically added by Babel.
- Transform source code.

In the following code, we show an example of a Babel-transformed code snippet:

```
// Babel Input: ES2015 arrow function
["Lion", "Tiger", "Shark"].map((animal) => console.
log(animal));


// Babel Output: ES5 equivalent
["Lion", "Tiger", "Shark"].map(function(animal) {
   console.log(animal)
});
```

You will notice that in the preceding code snippet, the modern arrow function is automatically transpiled to the function keyword that is supported by all browsers. This is what Babel does under the hood to your source code.

Next, let's understand where webpack comes in.

# Webpack

webpack is also a transpiler, and can perform the same function as Babel, and even more. webpack can package and bundle just about anything, including *images*, *HTML*, *CSS*, and *JavaScript*, into a single optimized script that can easily be used in the browser.

In this section, we'll leverage both Babel and webpack to show you how to set up a cross-platform JavaScript project. Let's dive right in.

## Example project using Babel and webpack

In this section, we're going to create a simple JavaScript project using npm. As such, you should have Node.js installed locally in order to follow along. Perform the following steps to achieve this:

1.  Open a terminal in your preferred directory and create a folder with the following commands:

    ```
    $ mkdir cross-env-js
    $ cd cross-env-js
    ```

    This will create a folder, cross-env-js, in your directory, and then change the directory as well.

2.  Create a package.json file. While you can do this manually, it is easier to create one using npm. Run the following command in your terminal:

    ```
    $ npm init -y
    ```

    The preceding code will create a package.json file and accept all default options. Ideally, this should output the following:

    ```
    {
      "name": "cross-env-js",
      "version": "1.0.0",
      "description": "",
      "main": "index.js",
      "scripts": {
        "test": "echo \"Error: no test specified\" && exit 1"
      },
      "keywords": [],
      "author": "",
      "license": "ISC"
    }
    ```

    Figure 1.1 – Output from running the npm init –y command

3.  Next, install all the relevant packages that will help us to perform bundling and transpilation:

    ```
    $ npm install --save-dev @babel/core @babel/cli @babel/
    preset-env babel-loader webpack webpack-cli
    $ npm install --save @babel/polyfill
    ```

Note that you install most packages as development dependencies. This is important so that your final code is not bundled with packages you only need during development. After installing these packages, your package.json file should look like this:

```
{
  "name": "cross-env-js",
  "version": "1.0.0",
  "description": "",
  "main": "index.js",
  "scripts": {
    "test": "echo \"Error: no test specified\" && exit 1"
  },
  "keywords": [],
  "author": "",
  "license": "ISC",
  "devDependencies": {
    "@babel/cli": "^7.12.8",
    "@babel/core": "^7.12.9",
    "@babel/preset-env": "^7.12.7",
    "babel-loader": "^8.2.2",
    "webpack": "^5.9.0",
    "webpack-cli": "^4.2.0"
  },
  "dependencies": {
    "@babel/polyfill": "^7.12.1"
  }
}
```

4.  Add some code, which we'll transpile and test. For this section, you can either create files and folders from the terminal or use a code editor. I'll use the Visual Studio Code editor here.

    In your code editor, open the cross-env-js project folder and then create the files and folders as follows:

```
├── dist
│   └── index.html
├── src
```

```
|    ├── index.js
|    ├── utils.js
```

That is, you will create two folders called `dist` and `src`. `dist` will contain an HTML file (`index.html`), which will be used to test our bundled application, and `src` will contain our modern JavaScript code that we want to transpile.

After creating these files and folders, your entire directory structure should look like this:

```
├── dist
|    └── index.html
├── node_modules
├── package-lock.json
├── package.json
└── src
     ├── index.js
     └── utils.js
```

> **Note**
>
> If you're using version control such as Git, you will typically add a `.gitignore` file to specify that `node_modules` can be ignored.

5.  Create a `dist` folder, and in that folder, create an `index.html` file with the following code:

```html
<!DOCTYPE html>
<html lang="en">
<head>
    <meta charset="UTF-8">
    <meta name="viewport" content="width=device-width,
initial-scale=1.0">
    <script src="bundle.js"></script>
    <title>Cross Environment Support</title>
</head>
<body>

</body>
</html>
```

The HTML file should be familiar to you, but notice that we added a `script` tag pointing to a `bundle.js` file. This file does not yet exist and will be generated by webpack using Babel under the hood.

6.  Write some modern JavaScript in the `src` folder. Starting with `utils.js`, we'll create and export some functions, and then import them to be used in `index.js`.

    Starting with `utils.js`, add the following code:

```
const subjects = {
    John: "English Language",
    Mabel: "Mathematics",
    Mary: "History",
    Joe: "Geography"
}


export const names = ["John", "Mabel", "Mary", "Joe"]
export const getSubject = (name) =>{
    return subjects[name]
}
```

    The `utils.js` script uses some modern JS syntax, such as `export` and arrow functions, and these will only be compatible with older browsers after transpilation.

    Next, in the `index.js` script, you'll import these functions and use them. Add the following code to your `index.js` script:

```
import { names, getSubject } from "./utils";
names.forEach((name) =>{
    console.log(`Teacher Name: ${name}, Teacher Subject:
${getSubject(name)}`)
})
```

    You'll notice that we are also using arrow functions and the destructuring import in the `index.js` file. Here, we're importing the exported array (names) and the `getSubject` function from the `utils.js` script. We are also using a combination of the arrow function and template literals (` `` `) to retrieve and log the details of each `Teacher`.

7.  Now that our modern JS files are ready, we'll create a configuration file that tells webpack where to find our source code to bundle as well as which transpiler to use, in our case, Babel.

In your root directory, create a `webpack.config.js` file and add the following code:

```js
const path = require('path');
module.exports = {
  entry: './src/index.js',
  output: {
    filename: 'bundle.js',
    path: path.resolve(__dirname, 'dist'),
    publicPath: '/dist'
  },
  module: {
    rules: [
      {
        test: /\.js$/,
        exclude: /(node_modules)/,
        use: {
          loader: 'babel-loader',
        }
      }
    ]
  }
};
```

Let's understand what is going on in this file:

a) The first part of the config file requires the `path` module, which will help resolve all path-related functions.

b) Next, you will notice the `entry` field. This field simply tells webpack where to find the starting/main script. webpack will use this file as a starting point, and then recursively walk through each import dependency to link all files relating to the entry file.

c) The next field is `output`, and this tells webpack where to save the bundled file. In our example, we are saving bundled files to the `dist` folder under the name `bundle.js` (remember we referenced `bundle.js` in our HTML file).

d) Finally, in the `module` field, we specify that we want to transpile each script using Babel, and we also exclude transpiling `node_modules`. With this webpack configuration file, you're ready to transpile and bundle your source code.

8.  In your `package.json` file, you'll add a script command that will call `webpack`, as shown in the following code block:

```
{

    ...

    "scripts": {
      "build": "webpack --mode=production",
      "test": "echo \"Error: no test specified\" && exit 1"
    },


    ...

}
```

9.  In your terminal, run the following command:

```
$ npm run build
```

This command will call the build script in your `package.json` file, and this, in turn, will ask webpack to bundle your code referencing the config file you created earlier.

Following successful compilation, you should have the following output in your terminal:

```
↱ cross-env-js git:(main) x npm run build

> cross-env-js@1.0.0 build /Users/mac/Documents/Building-Data-Driven-Applications-with-Danfo.js-/Chapter02/cross-env-js
> webpack

[webpack-cli] Compilation finished
asset bundle.js 222 bytes [compared for emit] [minimized] (name: main)
orphan modules 230 bytes [orphan] 1 module
./src/index.js + 1 modules 380 bytes [built] [code generated]
webpack 5.9.0 compiled successfully in 583 ms
```

Figure 1.2 – webpack bundling successful output

Upon successful completion of the preceding steps, you can navigate to the `dist` folder where you will find an extra file – `bundle.js`. This file has already been referenced by the `index.html` file, and as such will be executed anytime we load the `index.html` file in the browser.

To test this, open the `index.html` file in your default browser. This can be done by navigating to the directory and double-clicking on the `index.html` file.

Once you have the `index.html` file opened in your browser, you should open the developer console, where you can find your code output, as in the following screenshot:

Figure 1.3 – index.js output in the browser console

This shows that you have successfully transpiled and bundled your modern JS code into a format that can be executed in any browser, be it old or new.

> **Further reading**
>
> Bundling files can quickly become difficult and confusing, especially as the project gets bigger. If you require further understanding of how to bundle files, you can reference the following resources:
>
> * Getting Started (`https://webpack.js.org/guides/getting-started/`) with webpack
>
> * Usage Guide (`https://babeljs.io/docs/en/usage`) for Babel
>
> * How to enable ES6 (and beyond) syntax with Node and Express (`https://www.freecodecamp.org/news/how-to-enable-es6-and-beyond-syntax-with-node-and-express-68d3e11fe1ab/`)

In the next section, you'll learn how to set up testing and perform unit testing in your JavaScript application.

# Unit testing with Mocha and Chai

Writing tests for your application code is very important, but rarely talked about in most books. This is why we have decided to add this section on unit testing with Mocha. While you may not necessarily write verbose tests for every sample app you'll be building in this book, we will show you the basics you need to get by and you can even use them in your own project.

Testing, or automated testing, is used during development to check that our code actually behaves as expected. That is, you, the writer of a function, will generally know beforehand how the function behaves and therefore can test the outcome with the expected outcome.

**Mocha** is a popular and feature-rich test framework for JavaScript. It provides various testing functions, such as `it` and `describe`, which can be used to write and run tests automatically. The beautiful thing about Mocha is that it can run in both node and browser environments. Mocha also supports integration with various assertion libraries such as *Chai* (`https://www.chaijs.com/`), *Expect.js* (`https://github.com/LearnBoost/expect.js`), *Should.js* (`https://github.com/shouldjs/should.js`), or even Node.js' built-in *assert* (`https://nodejs.org/api/assert.html`) module. In this book, we'll use the Chai assertion library, as it is one of the most commonly used assertion libraries with Mocha.

## Setting up a test environment

Before we begin writing tests, we'll set up a basic Node.js project. Perform the following steps to achieve this:

1.  In your current working directory, create a new folder called `unit-testing`:

    ```
    $ mkdir unit-testing
    $ cd unit-testing
    ```

2.  Initialize a new Node.js project using npm, as shown in the following command:

    ```
    $ npm init -y
    ```

3.  Install Mocha and Chai as development dependencies:

    ```
    $ npm install mocha chai --save-dev
    ```

4.  Following successful installation, open your `package.json` file and change the `test` command in `scripts` to the following:

    ```
    {
        ...

        "scripts": {
            "test": "mocha"
        },
        ...
    }
    ```

    This means that we can run tests by running the `npm run test` command in the terminal.

5. Create two folders, `src` and `test`. The `src` folder will contain our source code/scripts, while the `test` folder will contain corresponding tests for our code. Your project tree should look like the following after creating the folders:

```
├── package-lock.json
├── package.json
├── src
└── test
```

6. In the `src` folder, create a script called `utils.js`, and add the following functions:

```javascript
exports.addTwoNumbers = function (num1, num2) {
    if (typeof num1 == "string" || typeof num2 == "string") {
        throw new Error("Cannot add string type to number")
    }
    return num1 + num2;
};
exports.mean = function (numArray) {
    let n = numArray.length;
    let sum = 0;
    numArray.forEach((num) => {
        sum += num;
    });
    return sum / n;
};
```

The preceding functions perform some basic computation. The first one adds two numbers and returns the result, while the second function computes the mean of numbers in an array.

> **Note**
> We are writing pre-ES16 JavaScript here. This is because we do not plan to set up any transpiler for this sample project. In a project using modern JavaScript, you'll typically transpile source code before testing it.

7. In your `test` folder, add a new file, also called `utils.js`. This naming convention is recommended, as different files should bear the same name as their corresponding source code. In the `utils.js` file in your `test` folder, add the following code:

```
const chai = require("chai");
const expect = chai.expect;
const utils = require("../src/utils");
describe("Test addition of two numbers", () => {
  it("should return 20 for addition of 15 and 5", () => {
    expect(utils.addTwoNumbers(15, 5)).equals(20);
  });

  it("should return -2 for addition of 10 and -12", () =>
{
    expect(utils.addTwoNumbers(10, -12)).equals(-2);
  });

  it("should throw an error when string data type is
passed", () => {
    expect(() => utils.addTwoNumbers("One", -12)).
to.throw(
      Error,
      "Cannot add string type to number"
    );
  });
});

describe("Test mean computation of an array", () => {
  it("should return 25 as mean of array [50, 25, 15,
10]", () => {
    expect(utils.mean([50, 25, 15, 10])).equals(25);
  });
  it("should return 2.2 as mean of array [5, 2, 1, 0,
3]", () => {
    expect(utils.mean([5, 2, 1, 0, 3])).equals(2.2);
  });
});
```

In the first three lines of the preceding code snippet, we are importing `chai` and `expect`, as well as the `utils` scripts, which contain our source code.

Next, we use Mocha's `describe` and `it` functions to define our test cases. Note that we have two `describe` functions corresponding to the two functions we have in our source code. This means that each `describe` function will contain individual unit tests that test different aspects of our code.

The first `describe` function tests the `addTwoNumber` function and includes three unit tests, one of which tests whether the correct error is thrown on a passing string data type. The second `describe` function tests the `mean` function by providing different values.

8.  To run our test, go to your terminal and run the following command:

```
$ npm test
```

This command executes the script test defined in your `package.json` file, and outputs a formatted test case report, as shown in the following screenshot:

```
→  unit-test git:(main) ✗ npm test

> unit-test@1.0.0 test /Users/mac/Documents/Building-Data-Driven-Applications-with-Danfo.js-/Chapter02/unit-test
> mocha


  Test addition of two numbers
    ✓ should return 20 for addition of 15 and 5
    ✓ should return -2 for addition of 10 and -12
    ✓ should throw an error when string data type is passed

  Test mean computation of an array
    ✓ should return 25 as mean of array [50, 25, 15, 10]
    ✓ should return 2.2 as mean of array [5, 2, 1, 0, 3]


  5 passing (12ms)
```

Figure 1.4 – Mocha test output showing all tests passed

By following the preceding steps, we were able to write and run some tests that passed on the first run. This may not be the case most times, as your test will generally fail a lot before passing, especially when you have numerous unit tests for different edge cases.

As an example, we'll add a new test case that expects an error when the array passed to the mean function contains no element.

In the test script, under the second `describe` function, add the following unit test:

```
...
  it("should throw error on empty array arg", () => {
      expect(() => utils.mean([])).to.throw(Error, "Cannot
compute mean of empty array")
```

```
});
...
```

By running the test again, we'll see the following error:

```
Test addition of two numbers
  ✓ should return 20 for addition of 15 and 5
  ✓ should return -2 for addition of 10 and -12
  ✓ should throw an error when string data type is passed

Test mean computation of an array
  ✓ should return 25 as mean of array [50, 25, 15, 10]
  ✓ should return 2.2 as mean of array [5, 2, 1, 0, 3]
  1) should throw error on empty array arg


5 passing (10ms)
1 failing

1) Test mean computation of an array
     should throw error on empty array arg:
   AssertionError: expected [Function] to throw Error
   at Context.<anonymous> (test/utils.js:31:42)
   at processImmediate (internal/timers.js:456:21)



npm ERR! Test failed.  See above for more details.
```

Figure 1.5 – Mocha test output showing one failed test

The error message provided by Mocha tells us that our function is expected to throw an error when an empty array is passed, but it is currently not doing that. To fix this error, we'll go to our source code and update the mean function, as shown in the following code block:

```
exports.mean = function (numArray) {
  if (numArray.length == 0){
    throw new Error("Cannot compute mean of empty array")
  }
  let n = numArray.length;
  let sum = 0;
  numArray.forEach((num) => {
    sum += num;
  });

  return sum / n;
};
```

Now, if we run the test again, we should see it pass successfully:

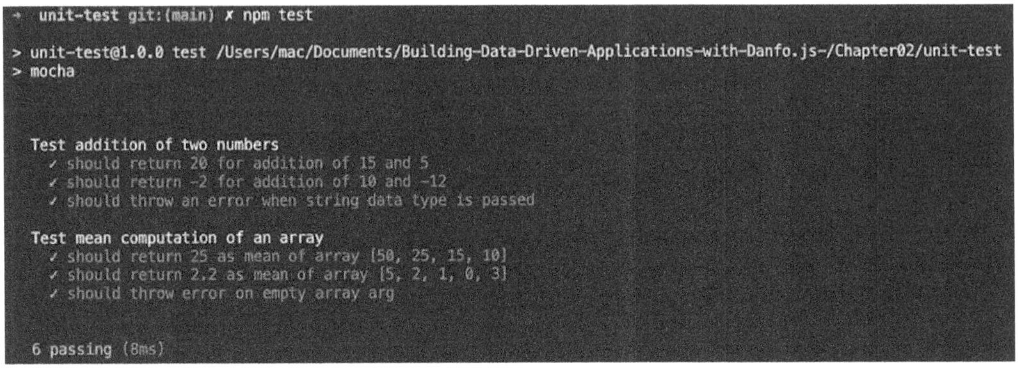

Figure 1.6 – Mocha test output showing that all tests passed

---

**Further reading**

Mocha is versatile and provides support for almost all test cases and scenarios you'll encounter. To learn more, you can visit the official documentation here: `https://mochajs.org/`.

Chai, on the other hand, provides numerous assertion statements and functions that you can use to enrich your test. You can learn more about these assertions here: `https://www.chaijs.com/api/`.

---

Congratulations on making it to the end of this chapter! It was a lengthy one, but the concepts covered are important, as they will help you build better data-driven products, as you'll see in future chapters.

# Summary

In this chapter, we introduced and discussed some of the modern JavaScript syntax introduced in ECMA 6. We first considered the difference between `let` and `var` and discussed why `let` is the preferred method for initializing variables. Following that, we discussed destructuring, the spread operator, scopes, and also closures. We then went on to introduce some important methods of arrays, objects, and strings. Following that, we talked about arrow functions, including their advantages over traditional functions, and then we moved on to discuss JavaScript promises and async/await.

We then looked briefly at OOP concepts and support in JavaScript with examples showing how to write classes. We also learned how to set up a modern JavaScript environment with transpiling and bundling support using tools such as Babel and webpack. Finally, we introduced unit testing using Mocha and the Chai library.

In the next chapter, we will introduce Dnotebook, an interactive computing environment that enables quick and interactive experimentation in JavaScript.

# Section 2: Data Analysis and Manipulation with Danfo.js and Dnotebook

This section introduces the reader to Danfo.js and Dnotebook (an interactive computing environment in JavaScript). It also dives deep into the internals of Danfo.js, examining DataFrames and Series, data transformation and analysis, plotting and visualization, and data aggregation and group operations.

This section comprises the following chapters:

- *Chapter 2, Dnotebook - An Interactive Computing Environment for JavaScript*
- *Chapter 3, Getting Started with Danfo.js*
- *Chapter 4, Data Analysis, Wrangling, and Transformation*
- *Chapter 5, Data Visualization with Plotly.js*
- *Chapter 6, Data Visualization with Danfo.js*
- *Chapter 7, Data Aggregation and Group Operations*

# 2
# Dnotebook - An Interactive Computing Environment for JavaScript

The idea of making our code expressive enough for humans to read and not just for machines to consume was well pioneered by Donald Knuth, who also wrote a book about it called *Literate Programming* (https://www.amazon.com/Literate-Programming-byKnuth-Knuth/dp/B004WKFC4S). Tools such as Jupyter Notebook give equal weight to prose and code, hence programmers and researchers are free to express themselves extensively through code and text (including images and workflows).

In this chapter, you're going to learn about **Dnotebook** – an interactive coding environment for JavaScript. You will learn how to install Dnotebook locally. You will also learn how to write code and Markdown in it. In addition, you will learn how to save and import the saved notebook.

The following topics will be covered in this chapter:

- Introduction to Dnotebook
- Setup and installation of Dnotebook
- Basic concepts behind interactive computing in Dnotebook
- Writing interactive code
- Working with Markdown cells
- Saving notebooks

# Technical requirements

To successfully follow along in this chapter, you need to have **Node.js** and a modern browser such as Chrome, Safari, Firefox, or Opera installed on your computer.

To install Node.js, you can follow the official guide here: `https://nodejs.org/en/`.

The code for this chapter is available and can be cloned from GitHub at `https://github.com/PacktPublishing/Building-Data-Driven-Applications-with-Danfo.js/tree/main/Chapter02`

# Introduction to Dnotebook

Over the past few years in the field of data science, interactive computing environments such as Jupyter Notebook and JupyterLab have actually made a huge impact in terms of how code is shared, and this has enhanced fast and rapid iterations of ideas.

In recent times, data science is moving towards the browser side, in order to support diverse users such as web developers. This means that a lot of mature data science tooling available in the Python ecosystem needed to be ported or made available in JavaScript as well. Following this line of reasoning, we, the authors of this book, as well as the creators of Danfo.js, decided to create a new version of the Jupyter Notebook specifically targeted at the JavaScript ecosystem.

Dnotebook, as we have called it, helps you to perform quick and interactive experimentation/prototyping in JavaScript. That means you can write code and view the results instantly in an interactive and notebook-like manner as seen in the following screenshot:

```
1 table(df2.head())
```

| | PassengerId | Survived | Pclass | Name | Sex | Age | SibSp | Parch | Ticket | Fare |
|---|---|---|---|---|---|---|---|---|---|---|
| 0 | 1 | 0 | 3 | Braund, Mr. Owen Harris | male | 22 | 1 | 0 | A/5 21171 | 7.25 |
| 1 | 2 | 1 | 1 | Cumings, Mrs. John Bradley (Florence Briggs Thayer) | female | 38 | 1 | 0 | PC 17599 | 71.2833 |
| 2 | 3 | 1 | 3 | Heikkinen, Miss. Laina | female | 26 | 0 | 0 | STON/O2. 3101282 | 7.925 |
| 3 | 4 | 1 | 1 | Futrelle, Mrs. Jacques Heath (Lily May Peel) | female | 35 | 1 | 0 | 113803 | 53.1 |
| 4 | 5 | 0 | 3 | Allen, Mr. William Henry | male | 35 | 0 | 0 | 373450 | 8.05 |

Understanding the numerical columns in the dataset.

```
1 table(df2.describe())
```

| | PassengerId | Survived | Pclass | SibSp | Parch | Fare |
|---|---|---|---|---|---|---|
| count | 891 | 891 | 891 | 891 | 891 | 891 |
| mean | 446.000031 | 0.383838 | 2.308642 | 0.523008 | 0.381594 | 32.204208 |
| std | 257.353842 | 0.486592 | 0.836071 | 1.102743 | 0.806057 | 49.693429 |
| min | 1 | 0 | 1 | 0 | 0 | 0 |
| median | 446 | 0 | 3 | 0 | 0 | 14.4542 |
| max | 891 | 1 | 3 | 8 | 6 | 512.329224 |
| variance | 66231 | 0.236772 | 0.699015 | 1.216043 | 0.649728 | 2469.436846 |

```
1 //Display the number of rows and columns
2 console.log(df2.shape)
```

[4990,13]

Figure 2.1 – Sample Interactive coding using Dnotebook

Dnotebook can be used in numerous areas and for different things such as the following:

- **Data science/analysis**: It can help you easily perform interactive data exploration and analysis using efficient JavaScript packages such as *Danfo.js*, *Plotly.js*, *Vega*, *Imagecook*, and so on.

- **Machine learning**: It can help you easily build, train, and prototype machine learning models using machine learning libraries such as *Tensorflow.js*.

- **Learning JavaScript interactively**: It can help you learn or teach JavaScript in an interactive and visual style. This can foster learning and understanding.

- **Plain experimentation/prototyping**: Any experimentation that can be written in JavaScript can run on Dnotebook, hence this can aid quick experimentation with ideas.

Now that you have an idea of what Dnotebook is, let's learn how to set up and use it locally.

# Setup and installation of Dnotebook

To get Dnotebook installed and running locally, you need to ensure that Node.js is installed. Once Node.js is installed, you can easily install Dnotebook by running the following command in your terminal:

```
npm install -g dnotebook
```

The preceding command installs Dnotebook globally. This is the recommended way of installing it, as it ensures that the Dnotebook server can be started from anywhere on our computer.

> **Note**
>
> You can also use Dnotebook online without installing it; check out the Dnotebook playground (`https://playnotebook.jsdata.org/ demo`).

After installation, you can start the server by running the following in a terminal/command prompt:

```
> dnotebook
```

This command will open up a tab as shown in the following screenshot in your default browser at port http://localhost:4400:

Figure 2.2 – Dnotebook home page

The opened page is the default page for the Dnotebook interface, and from here you can start writing JavaScript and Markdown.

> **Note**
>
> We are currently using **Dnotebook version 0.1.1**, and as such, when using this book in the future, you might notice some minor changes, especially in the UI.

# Basic concepts behind interactive computing in Dnotebook

In order to write interactive code/Markdown in Dnotebook, you need to understand some concepts such as cells and persistency/state. We start this section by explaining these concepts.

## Cells

A cell in Dnotebook is a unit block where code or text can be written in order to be executed. The following is an example screenshot showing code and Markdown cells:

Figure 2.3 – Empty code and Markdown cells in Dnotebook

Each cell has edit buttons that can be used for different purposes, as you'll see in the following screenshot:

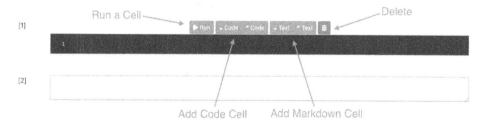

Figure 2.4 – Action buttons available in each cell

Now, let's understand what these buttons do:

- **Run**: The **Run** button can be used to execute a cell in order to show the output.
- **Add Code**: The add code button has two variants (up and down) specified by the arrow direction. They can be used to add a code cell above or below the current cell.
- **Add Markdown**: The add markdown button, like the add code button, has two variants to add Markdown cells either below or above the current cell.
- **Delete**: As the name suggests, this button can be used to delete a cell.

There are two types of cells, namely code cells and markdown cells.

## Code cells

A **code cell** is a cell where any JavaScript code can be written and executed. The first cell in a new notebook is always a code cell, and we can test this out with the classic hello world example.

In your open Dnotebook, write the following command and click the **Run** button:

```
console.log('Hello World!')
```

> **Note**
> Hovering over a code cell reveals the **Run** button. Alternatively, you can use the shortcut *Ctrl + Enter* in Windows or *Command + Enter* in Mac to run a code cell.

The hello world code and output should be similar to the following screenshot:

[1]
```
1 console.log('Hello World!')
```

Hello World!

Figure 2.5 – A code cell and executed output in Dnotebook

Next, let's understand Markdown cells.

# Markdown cells

**Markdown cells** are similar to code cells except for the fact that they can only execute Markdown or text. This means that Markdown text can compile any text written with Markdown syntax.

A Markdown cell in Dnotebook is typically white and can be opened by clicking the **Text** button in an open cell. The **Text** button is typically available for each cell, as shown in the following screenshot:

Figure 2.6 – Opening a Markdown cell in Dnotebook

Clicking on the **Text** button opens a Markdown cell, as shown in the following screenshot:

Figure 2.7 – Writing Markdown text in a Markdown cell

Here, you can write any Markdown-flavored text, and when executed, the result is compiled to text and shown in place of the Markdown cell as follows:

[2]

## Hello World

Figure 2.8 – Output of a Markdown cell

Now, let's talk about another important concept in interactive programming, called **persistence/state**.

# Persistence/state

Persistence or state in interactive computing is the ability of environment variables or data to outlive (persist) the cell that created it. This means variables declared/created in one cell are available to another irrespective of the position of the cell.

Each instance of a Dnotebook runs a persistent state, and variables declared in a cell without the `let` and `const` declarations are available to all cells.

---

**Note**

There are two main ways we encourage you to declare variables when working in Dnotebook.

Option 1 – Without a declaration keyword (preferred method):

```
food_price = 100
clothing_price = 200
total = food_price + clothing_price
```

Option 2 – With the `var` global keyword (this works but is not preferable):

```
var food_price = 100
var clothing_price = 200
var total = food_price + clothing_price
```

Using keywords such as `let` or `const` makes the variables inaccessible in a new cell.

---

To understand this better, let's declare some variables and try to access them in multiple cells created after or before:

1.  Create a new cell in your open notebook and add the following code:

```
num1 = 20
num2 = 35
sum = num1 + num2
console.log(sum)

//output 55
```

Run this code cell and you'll see the sum printed just below the cell, as seen in the following screenshot:

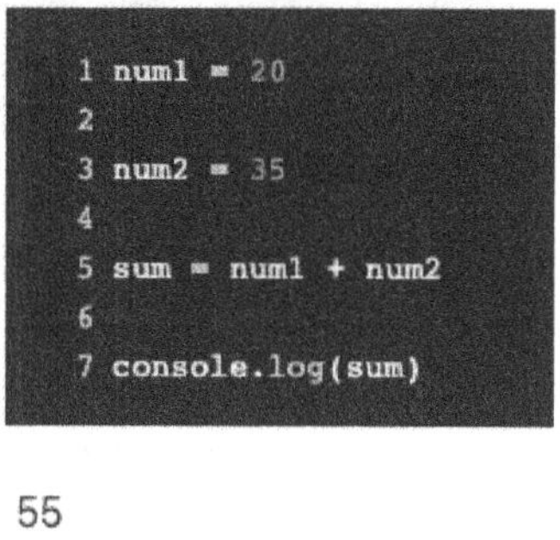

```
1 num1 = 20
2
3 num2 = 35
4
5 sum = num1 + num2
6
7 console.log(sum)
```

55

Figure 2.9 – Simple code to add two numbers

2.  Next, create a new cell following your first cell, by clicking on a code cell button, and try to use the sum variable as shown in the following code block:

```
newSum = sum + 30
console.log(newSum)

//outputs 85
```

By executing the preceding cell, you get the output of 85. This means that the variable sum from the first cell persists to the second cell as well as any other cell you'll create, as seen in the following screenshot:

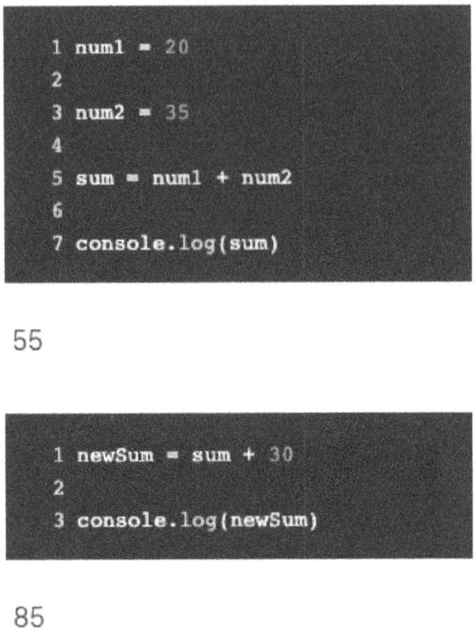

```
1 num1 = 20
2
3 num2 = 35
4
5 sum = num1 + num2
6
7 console.log(sum)
```

55

```
1 newSum = sum + 30
2
3 console.log(newSum)
```

85

Figure 2.10 – Two code cells sharing a persistent state

> **Note**
> Markdown cells do not persist variables, as they do not execute JavaScript code.

Now that you understand what cells and persistency are, you can now easily write interactive code in Dnotebook, and in the next section, we'll show you how to do that.

# Writing interactive code

In this section, we'll highlight some important things to know when writing interactive code in Dnotebook.

## Loading external packages

Importing external packages into your notebook is very important when writing JavaScript, and as such, Dnotebook has an inbuilt function called `load_package` for doing this.

The `load_package` method helps you to easily add external packages/libraries to your notebook via their CDN links. For instance, to load `Tensorflow.js` and `Plotly.js`, you can pass their CDN links to the `load_package` function, as shown in the following code:

```
load_package(["https://cdn.jsdelivr.net/npm/@tensorflow/
tfjs@3.4.0/dist/tf.min.js","https://cdn.plot.ly/plotly-latest.
min.js"])
```

This loads the packages and adds them to the notebook state so they can be accessed from any cell. In the following section, we use the `Plotly` library we just imported.

Add the following code to a new cell in your notebook:

```
trace1 = {
   x: [1, 2, 3, 4],
   y: [10, 11, 12, 13],
   mode: 'markers',
   marker: {
     size: [40, 60, 80, 100]
   }
};

data = [trace1];
layout = {
   title: 'Marker Size',
   showlegend: false,
   height: 600,
   width: 600
};

Plotly.newPlot(this_div(), data, layout); //this_div is a
built-in function that returns the current output's div name.
```

Executing the code cell in the preceding section will display a plot, as shown in the following screenshot:

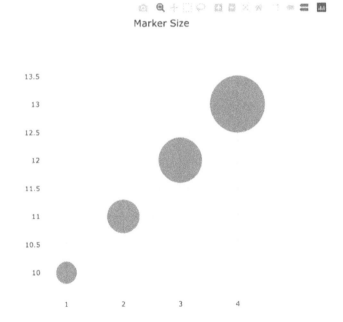

Figure 2.11 – Making a plot with an external package

Hence, by using `load_package`, you can add any external JavaScript package of your choice and work with it interactively in Dnotebook.

# Loading CSV files

Getting data into your notebook, especially into DataFrames, is very important. As such, another built-in function we introduce here is `load_csv`.

> **Note**
>
> DataFrames represent data in rows and columns. They are analogous to a spreadsheet or a database collection of rows and columns. We'll cover DataFrames and Series in depth in *Chapter 3, Getting Started with Danfo.js*.

The `load_csv` function helps you load CSV files over the internet into the `Danfo.js` DataFrame asynchronously. You should use this instead of Danfo's built-in `read_csv` function when reading big files and you want to track the progress. This is because `load_csv` displays a spinner on the navbar to indicate the progress.

Let's understand this better with an example. In a new code cell, add the following code:

```
load_csv("https://raw.githubusercontent.com/plotly/datasets/
master/finance-charts-apple.csv")
.then((data)=>{
  df = data
})
```

On executing the cell, if you look at the top-right corner, you'll notice a spinner that indicates the progress of the data loading.

After executing the cell, you can interact with the dataset as you would with a Danfo DataFrame. For instance, you can use another built-in function `table` to easily display the data in tabular format.

In a new cell, add the following code:

```
table(df)
```

On executing, you should see a table of your data, as shown in the following screenshot:

```
1 load_csv("https://raw.githubusercontent.com/plotly/datasets/master/finance-charts-apple.csv")
2 .then((data)=>{
3   df = data
4 })
```

```
1 table(df)
```

|   | Date | AAPL.Open | AAPL.High | AAPL.Low | AAPL.Close | AAPL.Volume | AAPL.Adjusted | dn |
|---|------|-----------|-----------|----------|------------|-------------|---------------|-----|
| 0 | 2015-02-17 | 127.489998 | 128.880005 | 126.919998 | 127.830002 | 63152400 | 122.905254 | 106.7410523 |
| 1 | 2015-02-18 | 127.629997 | 128.779999 | 127.449997 | 128.720001 | 44891700 | 123.760965 | 107.842423 |
| 2 | 2015-02-19 | 128.479996 | 129.029999 | 128.330002 | 128.449997 | 37362400 | 123.501363 | 108.8942449 |
| 3 | 2015-02-20 | 128.619995 | 129.5 | 128.050003 | 129.5 | 48948400 | 124.510914 | 109.7854494 |
| 4 | 2015-02-23 | 130.020004 | 133 | 129.660004 | 133 | 70974100 | 127.876074 | 110.3725162 |
| 5 | 2015-02-24 | 132.940002 | 133.600006 | 131.169998 | 132.169998 | 69228100 | 127.078049 | 111.0948689 |
| 6 | 2015-02-25 | 131.559998 | 131.600006 | 128.149994 | 128.789993 | 74711700 | 123.828261 | 113.2119183 |
| 7 | 2015-02-26 | 128.789993 | 130.869995 | 126.610001 | 130.419998 | 91287500 | 125.395469 | 114.1652991 |
| 8 | 2015-02-27 | 130 | 130.570007 | 128.240005 | 128.460007 | 62014800 | 123.510987 | 114.9668484 |

Figure 2.12 – Loading and displaying a CSV file

Next, we'll briefly look at another built-in function that aids the displaying of plots in your notebook.

## Getting a div container for plots

In order to display plots, most plotting libraries need some sort of container or HTML div. This is required for plotting using Danfo.js and the Plotly.js libraries. In order to make it easier to access an output div, Dnotebook comes with the this_div function built in.

The this_div function will return the HTML ID of the current code cell's output. For example, in the following code, we pass the this_div value to the plot method of a DataFrame:

```
const df = new dfd.DataFrame({col1: [1,2,3,4], col2:
[2,4,6,8]})

df.plot(this_div()).line({x: "col1", y: "col2"})
```

This will pass the current cell's div ID to the plot method of the DataFrame, and when executed, will display the resulting plot, as shown in the following screenshot:

```
1
2 const df = new dfd.DataFrame({col1: [1,2,3,4], col2: [2,4,6,8]})
3
4 df.plot(this_div()).line({x: "col1", y: "col2"})
```

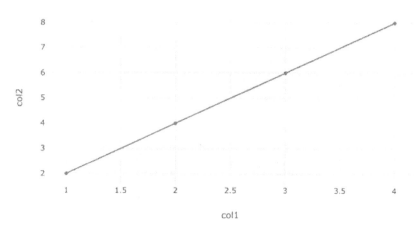

Figure 2.13 – Plotting a DataFrame

Finally, in the following section, we'll talk briefly about printing values inside a `for` loop. This does not work as expected, and we'll explain why.

## Gotchas when using a for loop

When you write a `for` loop and try to print each element in a Dnotebook code cell, you get the last element only. This issue has to do with the way the console works in the browser. For instance, try executing the following code and observe the output:

```
for(let i=0; i<10; i++){
    console.log(i)
}
//outputs 9
```

If you want to see all output when you run a `for` loop, especially when debugging in Dnotebook, you can use Dnotebook's built-in `forlog` method. This method has been appended to the default console object, and can be used as shown in the following code block:

```
for(let i=0; i<10; i++){
    console.forlog(i)
}
```

Executing the preceding code cell returns all values as shown in the following screenshot:

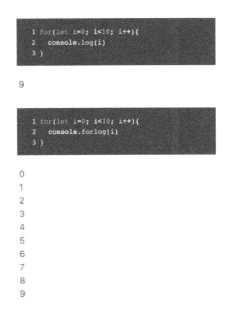

Figure 2.14 – Comparing the for and forlog methods

You'll notice that when using the `console.forlog` method, each output gets printed on a new line, just like the default behavior of `console.log` in a scripting environment.

In this section, we have covered some important functions and features that will be useful when writing interactive code in a Dnotebook environment. In the next section, we'll take a look at working with Markdown cells.

# Working with Markdown cells

Dnotebook supports Markdown, which gives the ability to mix your code with text and multimedia, hence enabling easy understanding for people who have access to the notebook.

Markdown is a markup language for creating formatted text using a plain-text editor. It is widely used in blogging, documentation pages, and README files. If you work with tools such as GitHub, then you have probably used Markdown.

Like many other tools, Dnotebook supports all Markdown syntax, image importing, the adding of links, and more.

In the following sections, we will see some important features you can leverage when using Markdown in Dnotebook.

## Creating a Markdown cell

In order to write Markdown in the Dnotebook environment, you need to add a Markdown cell by clicking the **Text** button (either up or down). This action adds a new Markdown cell to your notebook. The following screenshot shows example text written in the Markdown cell:

Fig 2.15 – Writing simple text in a Markdown cell

After writing Markdown text in a Markdown cell, you can click the **Run** button to execute it. This replaces the cell with the transpiled text in read mode. Double-clicking on the text reveals the Markdown cell again in edit mode.

## Adding images

In order to add images to Markdown cells, you can use the image syntax shown in the following snippet:

```
![alt Text](links to the image)
```

The following is the output:

Fig 2.16 – Adding images

For example, in the preceding screenshot, we add a link to an image available on the internet. The code is shown here:

```
![] (https://tinyurl.com/yygzqzrq)
```

The link provided is a link to a dog image. The **Run** button needs to be clicked to view the result of the image as shown here:

Fig 2.17 – Markdown image result

In the following sections, you'll learn some basic Markdown syntax that you can also add to your notebook. To see a comprehensive guide, you can visit the site `https://www.markdownguide.org/basic-syntax/`.

## Headings

To create a heading, you just need to add the pound symbol (`#`) in front of a word or phrase:

```
# First Heading
## Second Heading
### Third Heading
```

If we were to paste the previous texts into the Markdown and click the **Run** button, we would get the following output:

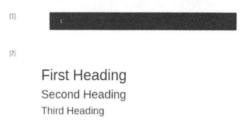

Fig 2.18 – Adding header text

In the result, you'll notice that a different number of pound signs in front of text results in a different size.

# Lists

Lists are important for enumerating objects and can be added by placing a star symbol (*) in front of text. We provide an example in the following section:

```
* Food
* Cat
    * kitten
* Dog
```

The preceding example creates an unordered list of the items **Food**, **Cat**, and **Dog**, with **kitten** as a sub-list of **Cat**.

In order to create a number list, just add numbers in front of the text, as shown here:

1. **First item**
2. **Second Item**
3. **Lot more**

The preceding text when input into the Markdown input field should output the following:

Fig 2.19 – List

In the following section, we will cover an important part of Dnotebook – saving. This is extremely important in order to reuse and share your notebooks with other people.

# Saving notebooks

Dnotebook supports the saving and importing of saved notebooks. Saving and importing a notebook gives you/others the ability to reuse your notebooks.

To save and import a notebook, click on the **File** menu, and select either the **Download Notebook** or **Upload Notebook** button depending on what you want to do. The options are displayed in the following screenshot:

Fig 2.20 – Saving and importing a notebook

Clicking **Download Notebook** saves the notebook in JSON format, and this can be easily shared or reloaded.

> **Saving and importing**
>
> To test this feature, go to `https://playnotebook.jsdata.org/`
> `demo`. Try saving the demo notebook. Then open up a new notebook,
> `https://playnotebook.jsdata.org`, and import the saved file.

# Summary

In this chapter, we introduced Dnotebook, an interactive library that supports text and multimedia. First, we covered the installation of Dnotebook locally and also pointed out that you can run a deployed version for free online. Next, we introduced some base concepts and gotchas when working with code and Markdown, and finally, we showed you how to save notebooks for sharing and reuse.

In the next chapter, we will get started with Danfo.js and introduce some of the underlying concepts of this awesome library.

# 3
# Getting Started with Danfo.js

The **Python** data science and machine learning ecosystem is very mature when compared to other programming languages, but when it comes to data presentation and the client side, **JavaScript** reigns supreme. From its robust data presentation tools to its ready availability in the browser to ease of setup, JavaScript is always recommended.

In this chapter, we will introduce you to the Danfo.js library, thus giving you an efficient tool for data analysis and manipulation in JavaScript. We'll cover the core data structures of Danfo.js – DataFrames and Series. We'll show you how to load different types of files, such as JSON, Excel, and CSV, into Danfo.js, and finally, you'll learn about some important functions available in Danfo.js that make data analysis and preprocessing in JavaScript easier.

In this chapter, we will cover the following topics:

- Why you need Danfo.js
- Installing Danfo.js
- Introducing Series and DataFrames
- Essential functions and methods in Danfo.js
- Data loading and working with different file formats

# Technical requirements

In order to follow along with this chapter, you should have the following:

- A modern browser such as Chrome, Safari, Opera, or Firefox
- **Node.js**, Danfo.js, and **Dnotebook** installed on your system

The code for this chapter is available here: `https://github.com/ PacktPublishing/Building-Data-Driven-Applications-with-Danfo. js/tree/main/Chapter03`.

> **Note**
> For most of the code snippets, you can make use of Dnotebook available online
> here: `https://playnotebook.jsdata.org/`.

# Why you need Danfo.js

To successfully bring a machine learning project written in Python to the web, there are a lot of processes and tasks to be carried out, things such as model deployment, creating API routes with frameworks such as **Flask**, **FastAPI**, or **Django**, and then using JavaScript to send HTTP requests to the model. You can clearly observe that the process involves a lot of JavaScript. It would be super awesome if we could perform all these processes in just JavaScript, wouldn't it? Well, the good news is that we can.

Over the past years, browsers have steadily increased in computational power and can support highly intensive computation, hence giving JavaScript the edge to challenge Python when it comes to data-intensive tasks.

With the help of Node.js, JavaScript has access to the GPU available on local computers, giving us the ability to undergo a full-stack machine learning project using Node.js for the backend and pure JavaScript for the frontend.

One of the benefits of using JavaScript is that inference can easily be done on the client side, hence data does not have to leave the client side to the server. Also, JavaScript gives our program cross-platform support and with the help of a **Content Delivery Network** (**CDN**), all the JavaScript packages needed for app development can easily be used without installing them.

With the introduction of tools such as **TensorFlow.js** and Danfo.js, we see more support for JavaScript in the data science and machine learning ecosystem.

Imagine a powerful data processing library such as **Python pandas** (`https://pandas.pydata.org/`) on the web, and think about the capability of infusing such a tool into popular modern frameworks in JavaScript such as **Vue** or **React**; the possibilities are endless. This is the power Danfo.js brings to the web.

There have been various attempts by JavaScript developers to bring the data processing capabilities of pandas to the web. Hence, we have libraries such as `pandas-js`, `dataframe-js`, `data-forge`, and `jsdataframe`.

The need to port Python pandas to JavaScript arose mainly because there was a need to perform data preprocessing and manipulation tasks in the browser.

This is a thread on **Stack Overflow** (`https://stackoverflow.com/questions/30610675/python-pandas-equivalent-in-JavaScript/43825646`) that gives a detailed overview of different tools for data preprocessing and manipulation in JavaScript, but most of these tools failed because of the following two reasons as explained in the thread:

- Lack of collaboration into a single library (lots of libraries trying to do different things)

- Most of the libraries not having the main features of pandas, as shown: `https://github.com/pandas-dev/pandas#main-features`

Besides the two reasons listed here, another issue we have observed when trying most of the existing tools is the lack of a good user experience like Python's pandas. pandas is very popular, and since most of the tools created in JavaScript mimic pandas, it is better to have a user experience that is quite similar to pandas.

Danfo.js was built to cover the gaps and issues faced by existing data processing libraries. The API has been carefully modeled after the pandas API, and as such provides a similar experience for people coming from a Python background, while at the same time providing a simple and intuitive API for JavaScript developers.

Besides a simple and familiar API, Danfo.js is faster thanks to TensorFlow.js, which powers all mathematical operations. Danfo.js has gained popularity in the JavaScript data science community with over 1,500 stars on GitHub in less than a year, as well as acceptance by the TensorFlow.js team at Google (`https://blog.tensorflow.org/2020/08/introducing-danfo-js-pandas-like-library-in-JavaScript.html`). Also, it is worth mentioning that Danfo.js is well maintained and under constant development thanks to active contributors, who ensure it is up to date and stable.

Apart from the preceding reasons, there are many more reasons why we decided to write this book on Danfo.js so as to give you the required skills and knowledge to perform data manipulation tasks in JavaScript. In the next section, we will start by learning how to install and import Danfo.js both in the browser and also in a Node.js environment.

# Installing Danfo.js

Danfo.js is readily available in both the browser and a Node.js environment. To use Danfo.js in the browser, you can add the `script` tag to the header of your HTML file as follows:

```
<head>
    ...
    <script src="https://cdn.jsdelivr.net/npm/danfojs@0.2.7/
lib/bundle.min.js"></script>
    ...
</head>
```

At the time of writing, the latest version of Danfo.js for a browser environment is 0.2.7. This will most likely have changed, but rest assured that all code and snippets used in this book will work in future versions.

> **Note**
>
> To install or get the latest version of Danfo.js, you can check the release page here, `https://danfo.jsdata.org/release-notes`, in the official documentation.

In a Node.js environment, Danfo.js can be installed using **Node package manager** (**npm**) or `yarn`, as shown in the following code:

```
//NPM
npm install danfojs-node
// Yarn
yarn add danfojs-node
```

Once installed, you can import Danfo.js using any of the following commands:

```
//Common js style
const dfd = require("danfojs-node")
```

```
// ES6 style
import * as dfd from 'danfojs-node'
```

> **Note**
>
> For practice, you can always make use of `https://playnotebook.jsdata.org/`, as discussed in the previous chapter. With this, you have online access to make use of Dnotebook for fast experimentation. To make use of the latest version of Danfo.js in Dnotebook, you can always use the `load_package` function.

The browser and Node.js versions of Danfo.js follow the same API design. The major difference is that in the browser, Danfo.js is added to the global scope under the name `dfd`. Hence, the `dfd` variable is available to all JavaScript or HTML files.

Now that we are done with the installation, we'll move on to the next section, where we discuss the major data structures and methods available in Danfo.js and Series.

# Introducing Series and DataFrames

Danfo.js exposes two main data structures, Series and DataFrames, to which all data manipulation can be done easily. DataFrames and Series provide a general representation for different types of data, hence the same data handling process can be applied to datasets with different formats.

In this section, we will look at different means of creating Series and DataFrames. We will see how to handle data in DataFrame and Series format. We will also look into different DataFrame and Series methods for data handling.

We will start this section by looking at how to handle data in a Series data structure.

## Series

A Series provides an entry point to handling one-dimensional data, such as a single array with a sequence of values of the same data type.

In this section, we will get familiar with Series data structures with the following Series methods:

```
*  table() and print() method:let sdata= new dfd.
Series([1,3,5,7,9,11])
table( sdata) // Displays the table in Dnotebook
```

The preceding code gives us the following output table:

|   | 0 |
|---|---|
| 0 | 1 |
| 1 | 3 |
| 2 | 5 |
| 3 | 7 |
| 4 | 9 |
| 5 | 11 |

Figure 3.1 – Series table

In the preceding code snippet, we make use of `table()` to print the Series table. This is because we are making use of Dnotebook for most of the code in this chapter. To print DataFrames or Series in a browser or Node.js console, you can always call `print()` as follows:

```
sdata.print() // will give us same output as table( sdata)
```

In subsequent code, we will always make use of the `table()` function, which allows us to print the DataFrame on the browser web page.

- The `.index` attribute: By default, the column name for a Series table is always 0 unless it is specified. Also note that the index of a Series table ranges from 0 to n – 1 of the data, where n is the length of the data. The index can also be obtained via the following:

```
console.log(sdata.index) // [0,1,2,3,4,5]
```

- `.dtype` and `astype()`: Sometimes the data passed into a Series might not be homogeneous, and may contain fields with mixed data types, as follows:

```
let sdata = new dfd.Series([1,2,3, "two:","three"])
console.log(sdata.dtype)
// string
```

`dtype` outputted `string`, whereas the data is of a different data type, hence we can always change `dtype` as shown in the following code:

```
sdata.astype('int32')
```

The preceding code changes the data type of the Series, but it does not actually cast the strings in the data to `int32`. Hence, this will throw an error if we perform a numerical operation on the data.

- The `.tensor` attribute: The data in a Series can be obtained in two main formats – as JavaScript **arrays** or **tensors**.

In the following code, we demonstrate how to obtain Series data as a JavaScript array and as a tensor:

```
sdata = new dfd.Series([1,2,3,4,5])console.log(sdata.
values)
// [1,2,3,4,5]
console.log(sdata.tensor)
// output Tensorflow.js tensor
```

The output from `series.tensor` is a valid TensorFlow.js tensor, and as such all supported tensor operations can be performed on it. For example, in the following code, we call the exponential function on the Series `tensor` object:

```
sdata.tensor.exp(2).arraySync()
// [3,7,20,55,148]
```

In the preceding code, we were able to print out the exponential of the Series data because we had access to the underlying tensor. The `arraySync` method returns the array format of a tensor. Let's look at the `set_index()` method.

- The `set_index()` method: The index can be specified when creating a Series data structure. We demonstrate this in the following code:

```
series = new dfd.Series([1,2,3,4,5], {index:
["one","two", "three", "four", "five"]})
table(series)
```

This sets the index of the Series and replaces the default numerical index as shown here:

| | 0 |
|---|---|
| one | 1 |
| two | 2 |
| three | 3 |
| four | 4 |
| five | 5 |

Figure 3.2 – Series with a named index

We can always change the default index value of a Series, as shown in the following code:

```
sdata = new dfd.
Series(["Humans","Life","Meaning","Fact","Truth"])
new_series = sdata.set_index({ "index": ["H", "L",
"M","F","T"] })
table(new_series)
```

`sdata.set_index()` resets the index and returns a new Series as shown in *Figure 3.2 (left)*, without mutating the original Series. We can set `set_index()` to actually update the index of the original Series and not return a new Series, by setting the `inplace` key to `true`:

```
sdata.set_index({ index: ["H", "L", "M","F","T"] ,
inplace: true })
```

We will now look at the `.apply()` method.

- The `.apply()` method: Since a Series is a single-dimensional array, it is easy to apply operations similar to how it would be done if working with an array in JavaScript. Series have a method called `.apply` to which you can apply any function on each value of the Series data:

```
sdata = new dfd.
Series(["Humans","Life","Meaning","Fact","Truth"])
series_new = sdata.apply((x) => {
  return x.toLocaleUpperCase()
})
table(series_new)
```

This converts each of the strings in the Series to uppercase by applying `x.toLocaleUpperCase()` on each of the values. The following figure shows the table output before and after applying `x.toLocaleUpperCase` on the Series data:

|   | 0 |
|---|---|
| **H** | Humans |
| **L** | Life |
| **M** | Meaning |
| **F** | Fact |
| **T** | Truth |

|   | 0 |
|---|---|
| **0** | HUMANS |
| **1** | LIFE |
| **2** | MEANING |
| **3** | FACT |
| **4** | TRUTH |

Figure 3.3 – Left: setting the index to string. Right: using apply to convert all strings to uppercase

The `.apply` method is also handy for performing numerical operations on each value of the Series data:

```
sf = new dfd.Series([1, 2, 3, 4, 5, 6, 7, 8])
sf_new = sf.apply(Math.log)
table(sf_new)
```

We used the `.apply` function to find the log of each value in the Series. The same can also be done for all other mathematical operations.

The preceding code outputs the following table:

|   | 0 |
|---|---|
| **0** | 0 |
| **1** | 0.6931471805599453 |
| **2** | 1.0986122886681096 |
| **3** | 1.3862943611198906 |
| **4** | 1.6094379124341003 |
| **5** | 1.791759469228055 |
| **6** | 1.9459101490553132 |
| **7** | 2.0794415416798357 |

Figure 3.4 – Applying math ops (Math.log) to each value of a Series

The Series data structure also contains the .map method, similar to .apply. This method maps each value of the Series to another value:

```
sf = new dfd.Series([1,2,3,4])
map = { 1: "ok", 2: "okie", 3: "frit", 4: "gop" }
sf_new = sf.map(map)
table(sf_new)
```

The map accepts objects and functions as parameters. In the preceding code, we created an object called map that contains the values in the Series as keys and the values that each key should be mapped to.

The preceding code outputs the following table:

| | 0 |
|---|---|
| **0** | ok |
| **1** | okie |
| **2** | frit |
| **3** | gop |

Figure 3.5 – Using map on Series values

Next, we will look at the .isna() method.

• The .isna() method: The data we've been passing into the Series so far seems to be okay and contains no missing values or undefined values, but when working with real data, we will always have NaN or undefined values. We can always check whether such values exist in our Series:

```
sf_nan = new dfd.Series([1,2, NaN, 20, undefined, 100])
table(sf_nan)
```

As a test, we create a Series containing NaN and an undefined variable. We obtain the following output:

| | 0 |
|---|---|
| **0** | 1 |
| **1** | 2 |
| **2** | NaN |
| **3** | 20 |
| **4** | NaN |
| **5** | 100 |

Figure 3.6 – Table containing a Series frame with NaN values

Let's check the rows containing `NaN` and undefined values; for each of these rows, we output `true`, and if not, they are represented with `false` as shown in the following code:

```
table(sf_nan.isna())
```

We obtain the following output:

| | 0 |
|---|---|
| **0** | false |
| **1** | false |
| **2** | true |
| **3** | false |
| **4** | true |
| **5** | false |

Figure 3.7 – The isna table for the Series data

The `.add()` method is as follows.

- The `.add()` method: The Series data structure also supports element-wise operations (such as addition, subtraction, multiplication, and division) between Series, and after this operation, the indexes are automatically aligned:

```
sf1  = new dfd.Series([2,20,23,10,40,5])
sf2  = new dfd.Series([30,20,40,10,2,3])
sf_add = sf1.add(sf2)
table(sf_add)
```

We have the following output:

| | 0 |
|---|---|
| 0 | 32 |
| 1 | 40 |
| 2 | 63 |
| 3 | 20 |
| 4 | 42 |
| 5 | 8 |

Figure 3.8 – Adding two Series

The preceding code snippet adds the sf2 Series to sf1, and each of the values is added element-wise, per row.

Note also that a single value can be passed to sf1.add, which will add a constant value to each of the elements in the sf1 Series:

```
table(sf1.add(10))
```

The following is the output of the preceding code:

| | 0 |
|---|---|
| 0 | 12 |
| 1 | 30 |
| 2 | 33 |
| 3 | 20 |
| 4 | 50 |
| 5 | 15 |

Figure 3.9 – Adding a constant value to Series elements

The operation shown previously applies for all other mathematical operations as well as Series.

In the next sub-section, we'll look into the operation of DataFrames, how to create a DataFrame, and how to handle data with DataFrame operations.

# DataFrames

A DataFrame represents a collection of Series, although unlike Series, it is basically a two-dimensional representation of data (although it can also represent higher dimensions) containing multiple columns and rows.

A DataFrame can basically be constructed using JavaScript objects, as shown in the following code:

```
data = {
   artist: ['Drake','Rihanna','Gambino', 'Bellion',
'Paasenger'],
   rating: [5, 4.5, 4.0, 5, 5],
   dolar: [ '$1m', '$2m','$0.1m', '$2.5m','$3m']
}
df = new dfd. DataFrame (data)
table(df)
// print out the  DataFrame
```

The resulting DataFrame should look like this:

| | artist | rating | dolar |
|---|---|---|---|
| 0 | Drake | 5 | $1m |
| 1 | Rihanna | 4.5 | $2m |
| 2 | Gambino | 4 | $0.1m |
| 3 | Bellion | 5 | $2.5m |
| 4 | Paasenger | 5 | $3m |

Figure 3.10 – DataFrame table

In the preceding code, JavaScript object data is created containing keys and values. Note the following about the JavaScript object:

- Each key contains a list value representing data per column.
- Each key represents the column and the column name.
- Each key must have the same length of list values.

The following are common methods for handling DataFrame data structures:

- The `head()` and `tail()` methods: Sometimes, while working on a big dataset, you might need to see the first set of rows (maybe the first 5 rows or the first 30 rows, as much as you like) and also the last rows in the data.

  Since our DataFrame contains five rows, let's print the first two rows of the table:

  ```
  table(df.head(2))
  ```

  The `head()` method by default prints out the first five rows of a DataFrame, but we can specify 2 since our data is small.

  The following table shows the result of the preceding code:

  |   | artist | rating | dolar |
  |---|--------|--------|-------|
  | 0 | Drake | 5 | $1m |
  | 1 | Rihanna | 4.5 | $2m |

  Figure 3.11 – Table for df.head(2)

  Also, to see the last set of rows in the DataFrame, we make use of the `tail()` method. By default, the `tail` method prints out the last five rows in the DataFrame:

  ```
  table(df.tail(2))
  ```

  The `tail` function takes in a value that specifies the number of rows to print from the end of the DataFrame table, as shown:

  |   | artist | rating | dolar |
  |---|--------|--------|-------|
  | 3 | Bellion | 5 | $2.5m |
  | 4 | Paasenger | 5 | $3m |

  Figure 3.12 – Printing the last two rows

- **Setting and getting column data**: Besides using JavaScript objects to specify the data to be passed into the DataFrame, we can pass in a list of values and column names into the DataFrame, as follows:

  ```
  data = [
      ['Drake', 5, '$1m'],
      ['Rihanna', 4.5, '$2m'],
      ['Gambino', 4.0, '$0.1m'],
      ['Bellion', 5, '$2.5m'],
  ```

```
    ['Passenger', 5, '$3m']
]
```

```
columns = ['artist', 'rating', 'dollar']
```

```
df = new dfd. DataFrame (data, {columns: columns})
```

The `data` variable contains the data, which is represented as an array of an array. Each array in the data array contains values arranged row-wise as it will be represented in the DataFrame table; [Drake, 5, $1m] represents a row in the DataFrame, while each value (Drake, 5, $1m) represents different columns. `Drake` belongs to the `artist` column and `$1m` belongs to the `dollar` column, as shown in the following figure:

|   | artist | rating | dollar |
|---|--------|--------|--------|
| **0** | Drake | 5 | $1m |
| **1** | Rihanna | 4.5 | $2m |
| **2** | Gambino | 4 | $0.1m |
| **3** | Bellion | 5 | $2.5m |
| **4** | Passenger | 5 | $3m |

Figure 3.13 – DataFrame table

Recall that we said a DataFrame is a collection of Series, hence it is possible to access each of these Series from the DataFrame like the way we access JavaScript object elements. Here is how to do that:

```
table(df.artist)
```

Note that `artist` is the name of a column in the DataFrame, hence if we were to think of a DataFrame as an object, `artist` is the key and the value of the key is a Series. The preceding code should output the following table:

| | artist |
|---|---|
| **0** | Drake |
| **1** | Rihanna |
| **2** | Gambino |
| **3** | Bellion |
| **4** | Passenger |

Figure 3.14 – Column Series

The `artist` column can also be accessed via `df['artist']`:

```
table(df['dollar'])
```

This method also outputs the same result as the initial `df.dollar` method being used:

| | dollar |
|---|---|
| **0** | $1m |
| **1** | $2m |
| **2** | $0.1m |
| **3** | $2.5m |
| **4** | $3m |

Figure 3.15 – Accessing column values

Apart from accessing column values via an object-like method, we can also change the column data as follows:

```
df['dollar'] = ['$20m','$10m', '$40m', '$35m', '$10m']
```

If we were to print out df.dollar, we would obtain the following output:

| | dollar |
|---|---|
| 0 | $20m |
| 1 | $10m |
| 2 | $40m |
| 3 | $35m |
| 4 | $10m |

Figure 3.16 – Changing the column value

The changing of column data can also be done by creating a Series frame and then assigning it to a column in the DataFrame, as follows:

```
rating = new dfd.Series([4.9,5.0, 7.0, 3.0,2.0])
df['rating'] = rating
table(df['rating'])
```

The preceding code changes the value of the column rating and outputs the following table:

| | rating |
|---|---|
| 0 | 4.9 |
| 1 | 5 |
| 2 | 7 |
| 3 | 3 |
| 4 | 2 |

Figure 3.17 – Changing column data with a Series value

This method of using Series data to update column data comes in handy when working on a time-related column. We might need to extract hours or months in a date column and replace the column with the extracted information (hours or months).

Let's add a date column to the DataFrame:

```
date = [ "06-30-02019", "07-29-2019", "08-28-2019",
"09-12-2019","12-03-2019" ]
df.addColumn({column: "date", value: date})
```

We make use of the addColumn method to add a new column to the DataFrame. The addColumn method accepts an object argument that must contain the following keys:

a) The name of the column to add

b) The new value for the column

The preceding code should output the following table:

| | artist | rating | dollar | date |
|---|---|---|---|---|
| 0 | Drake | 4.9 | $20m | 06-30-02019 |
| 1 | Rihanna | 5 | $10m | 07-29-2019 |
| 2 | Gambino | 7 | $40m | 08-28-2019 |
| 3 | Bellion | 3 | $35m | 09-12-2019 |
| 4 | Passenger | 2 | $10m | 12-03-2019 |

Figure 3.18 – Adding a new column

The new column created is a date-time column, hence we can convert this column to a time series data structure and then extract the month name for each data point in the column:

```
date_series = df['date']
df['date'] = date_series.dt.month_name()
```

We extract the date column and assign it to a date_series variable. From date_series (which is actually a Series data structure), we extract the month name for each value. A Series data structure contains the dt method, which converts the Series structure to a time series structure and then extracts the month name using the month_name() method.

The preceding code outputs the following table:

| | artist | rating | dollar | date |
|---|---|---|---|---|
| 0 | Drake | 4.9 | $20m | Jun |
| 1 | Rihanna | 5 | $10m | Jul |
| 2 | Gambino | 7 | $40m | Aug |
| 3 | Bellion | 3 | $35m | Sep |
| 4 | Passenger | 2 | $10m | Dec |

Figure 3.19 – Extracting the month name

The next methods that we will look into are `.values` and `.tensor`.

- `.values` and `.tensor`: The DataFrame values, that is, each column value, can be obtained as a giant array using `df.values`:

```
df.values
//output
[[Drake,4.9,$20m,Jun], [Rihanna,5,$10m,Jul], [Gambino,
7,$40m,Aug], [Bellion,3,$35m,Sep], [Passenger,2,$10m,Dec]]
```

Sometimes, we might want to obtain the DataFrame as a tensor for machine learning operations, hence the following shows how to obtain a DataFrame as tensor values:

```
df.tensor
//output
{"kept":false,"isDisposedInternal":false,"shape":[5,4],
"dtype":"string","size":20,"strides":[4],"dataId":{},
"id":28,"rankType":"2"}
```

The `tensor` attribute returned a TensorFlow.js tensor data structure. Since a tensor must always contain a unique data type, all the data points in our DataFrame are converted to a `dtype` string.

- The `.transpose()` method: Since we made the claim that a DataFrame is a two-dimensional representation of data, we should be able to transpose the DataFrame:

```
df.transpose()
```

`df.transpose()` moves the column list as an index and the index as a column, as shown in the following table:

| | 0 | 1 | 2 | 3 | 4 |
|---|---|---|---|---|---|
| **artist** | Drake | Rihanna | Gambino | Bellion | Passenger |
| **rating** | 4.9 | 5 | 7 | 3 | 2 |
| **dollar** | $20m | $10m | $40m | $35m | $10m |
| **date** | Jun | Jul | Aug | Sep | Dec |

Figure 3.20 – Transposing a DataFrame

We can obtain the index of the DataFrame and see whether the column is now the current index:

```
df2 = df.transpose()
```

```
console.log(df2.index)
// [artist,rating,dollar,date]
```

The preceding code is the index of the transposed DataFrame. In the preceding examples, the index has always been a numerical value, but with this, it shows that the index can also be a string.

- The `.set_index()` method: Using the DataFrame itself, before transposing it, let's change the index of the DataFrame from integers to a string index:

```
df3 = df.set_index({key:["a","b","c","d","e"]})
```

Using the `set_index` method in the DataFrame sets the index with the given key values. `set_index` also accepts the `inplace` key (the `inplace` key is a Boolean key it receives either `true` or `false` as a value, but the default value is `false`) key to specify whether the DataFrame should be updated or a new DataFrame containing the update should be returned.

The following table shows the result of the preceding code:

|   | artist | rating | dollar | date |
|---|--------|--------|--------|------|
| a | Drake | 4.9 | $20m | Jun |
| b | Rihanna | 5 | $10m | Jul |
| c | Gambino | 7 | $40m | Aug |
| d | Bellion | 3 | $35m | Sep |
| e | Passenger | 2 | $10m | Dec |

Figure 3.21 – Setting the index of the DataFrame

Since the index can be anything, we can also set the index of the DataFrame to be one of the columns in the DataFrame:

```
df4 = df.set_index({key:"artist"})
```

This time around, `set_index` takes in a value instead of an array of proposed index values. If a single value is passed as a key index, `set_index` believes it should be available as a column in the DataFrame.

The values in the `artist` column are used to replace the default index. Also note that the default value of `inplace` is `false`, hence a new DataFrame is created and returned with the updated index.

The following shows the output of the DataFrame created:

| | rating | dollar | date |
|---|---|---|---|
| Drake | 4.9 | $20m | Jun |
| Rihanna | 5 | $10m | Jul |
| Gambino | 7 | $40m | Aug |
| Bellion | 3 | $35m | Sep |
| Passenger | 2 | $10m | Dec |

Figure 3.22 – Setting the index to the artist column

By now, you should be aware of how to create a Series and DataFrame and how to make use of their respective methods for data processing.

In the next section, we look into some essential functions and methods in Danfo.js that will commonly be used for data analysis and processing.

# Essential functions and methods in Danfo.js

In this section, we will look at some important functions and methods in relation to Series and DataFrames. The methods in each of the data structures are a lot and we can always visit the documentation for more methods. This section will only mention some of the most commonly used methods:

- `loc` and `iloc` indexing
- Sorting: the `sort_values` and `sort_index` methods
- `Filter`
- Arithmetic operations such as `add`, `sub`, `div`, `mul`, and `cumsum`

## loc and iloc indexing

Accessing DataFrame rows and columns is made easier with the `loc` and `iloc` methods; both methods allow you to specify the rows and columns you would like to access. For those of you coming from Python' pandas library, the `loc` and `iloc` format as implemented in Danfo.js should be familiar.

The `loc` method is used to access a DataFrame with an index that is not numeric, as follows:

```
df_index = df4.loc({rows:['Bellion','Rihanna','Drake']})
```

Making use of the last updated DataFrame from the previous section, we used the `loc` method to obtain a certain row index from the DataFrame. This index value is specified in the `rows` key element:

| | rating | dollar | date |
|---|---|---|---|
| **Drake** | 4.9 | $20m | Jun |
| **Rihanna** | 5 | $10m | Jul |
| **Bellion** | 3 | $35m | Sep |

Figure 3.23 – Accessing specific row index values

Also, the `loc` and `iloc` methods give us the ability to slice an array. This should be familiar to Python users. Don't worry if you don't know what it is; the following examples will show what it means and how it is done.

To obtain a range of indexes, we can specify the upper and lower bounds even for a non-numerical index:

```
df_index = df4.loc({rows:["Rihanna:Bellion"]})
```

The `loc` method is used whenever the column name and index of a DataFrame is a string. For the preceding code, we specify that we want to extract data values between the `Rihanna` row index and the `Bellion` row index.

This colon helps specify the range boundary. The following table shows the output of the preceding code:

| | rating | dollar | date |
|---|---|---|---|
| **Rihanna** | 5 | $10m | Jul |
| **Gambino** | 7 | $40m | Aug |

Figure 3.24 – String indexing

Apart from indexing the rows, we can also extract a specific column:

```
df_index = df4.loc({rows:["Rihanna:Bellion"],
columns:["rating"]})
```

A `columns` key is passed into `loc` to extract the column called `rating`.

The following table shows the result of the preceding code:

| | rating |
|---|---|
| **Rihanna** | 5 |
| **Gambino** | 7 |

Figure 3.25 – Column and row indexing

We can also specify the range for the indexing column, as follows:

```
df_loc= df4.loc({rows:["Rihanna:Bellion"],
columns:["rating:date"]})
```

This should return a table similar to the following:

| | rating | dollar |
|---|---|---|
| **Rihanna** | 5 | $10m |
| **Gambino** | 7 | $40m |

Figure 3.26 – Column range

From the preceding table, we can see the range is quite different from the row indexing; the column indexing does exclude the upper bound.

The `loc` method in a DataFrame uses characters to index both the row data and the column data. `iloc` does the same but makes use of integers for indexing rows and columns.

Using the initial DataFrame as shown in the following table, we are going to perform indexing using the `iloc` method:

| | artist | rating | dollar |
|---|---|---|---|
| **0** | Drake | 5 | $20m |
| **1** | Rihanna | 4.5 | $10m |
| **2** | Gambino | 4 | $40m |
| **3** | Bellion | 5 | $35m |
| **4** | Passenger | 5 | $10m |

Figure 3.27 – The artist table

Let's use `iloc` to access index 2 to 4 of the table shown in *Figure 3.27*:

```
t_df = df.iloc({rows:['2:4']})
table(t_df)
```

The `iloc` method accepts the same keywords as the `loc` method: the `rows` and `columns` keywords. The numbers are wrapped in a string because we want to perform **slicing** of an array (fancy indexing), just as it was done in the `loc` method.

The following table shows the result of the preceding code:

| | artist | rating | dollar |
|---|---|---|---|
| 2 | Gambino | 4 | $40m |
| 3 | Bellion | 5 | $35m |

Figure 3.28 – Indexing table rows from 2 to 4 using a closed upper bound

---

**Slicing arrays (fancy indexing)**

Slicing arrays will be familiar to those coming from Python. In Python, it is possible to index an array using a range of values, for example, `test = [1,2,3,4,5,6,7] test [2:5] => [3,4,5]`. JavaScript does not have this property, hence Danfo.js does this by passing the range as a string, and from the string, the upper and lower bounds are extracted.

---

The same slicing operation as done in the previous example for rows can also be done on columns:

```
column_df = df.iloc({columns:['1:']})
table(column_df)
```

`df.iloc({columns:['1:']})` is used to extract the column starting from index 1 (which represents the `rating` column) up to the last column. For integer indexing, whenever the upper bound is not specified, it picks the last column as the upper bound.

In this section, we talked about indexing and different methods of indexing a DataFrame. In the next section, we will be talking about sorting a DataFrame by column value and row index.

# Sorting

Danfo.js supports two methods for sorting data – in terms of index or by a column value. Also, Series data can only be sorted by index, because it is a DataFrame object with a single column. DataFrames have a method called `sort_values` that enables you to sort data by a specific column, hence by sorting a particular column, we are sorting all other columns with respect to this column.

Let's create a DataFrame containing numbers:

```
data = {"A": [-20, 30, 47.3],
        "B": [ -4, 5, 6],
        "C": [ 2, 3, 30]}
df = new dfd. DataFrame (data)
table(df)
```

This outputs the following table:

|   | A | B | C |
|---|---|---|---|
| 0 | -20 | -4 | 2 |
| 1 | 30 | 5 | 3 |
| 2 | 47.3 | 6 | 30 |

Figure 3.29 – DataFrame of numbers

Now, let's sort the DataFrame by the values in column B:

```
df.sort_values({by: "C", inplace: true, ascending: false})
table(df)
```

The `sort_values` method takes in the following keyword arguments:

- by: The name of the column used to sort the DataFrame.
- `inplace`: Whether the original DataFrame should be updated or not (`true` or `false`).
- `ascending`: Whether the column should be sorted in ascending order or not. The default value is always `false`.

In the preceding code snippet, we specify to sort the DataFrame by column C and also set `inplace` to `true`, hence the DataFrame is updated and the column is also sorted in descending order. The following table shows the output:

|   | A | B | C |
|---|---|---|---|
| 2 | 47.3 | 6 | 30 |
| 1 | 30 | 5 | 3 |
| 0 | -20 | -4 | 2 |

Figure 3.30 – DataFrame sorted by column values

Sometimes we might not want to update the original DataFrame itself. Here lies the influence of the `inplace` keyword:

```
sort_df = df.sort_values({by: "C", inplace:false, ascending:
false})
table(sort_df)
```

In the preceding code, the DataFrame was sorted by a column. Let's try to sort the same DataFrame by its index:

```
index_df = sort_df.sort_index({ascending:true})
table(index_df)
```

`sort_index` also accepts keyword arguments: `ascending` and `inplace`. We set the `ascending` keyword to be `true`, which should give us the DataFrame with its index set in ascending order.

The following output is obtained:

|   | A | B | C |
|---|---|---|---|
| 0 | -20 | -4 | 2 |
| 1 | 30 | 5 | 3 |
| 2 | 47.3 | 6 | 30 |

Figure 3.31 – Sorted index DataFrame

Sorting in a Series requires quite a simple approach; sorting is done by calling the `sort_values` method, as follows:

```
data1 = [20, 30, 1, 2, 4, 57, 89, 0, 4]
series = new dfd.Series(data1)
sort_series = series.sort_values()
table(sort_series)
```

The `sort_values` method also accepts keyword arguments, `ascending` and `inplace`. By default, `ascending` is set to `true` and `inplace` is set to `false`.

The following table shows the result of the preceding code:

|   | 0 |
|---|---|
| 7 | 0 |
| 2 | 1 |
| 3 | 2 |
| 8 | 4 |
| 4 | 4 |
| 0 | 20 |
| 1 | 30 |
| 5 | 57 |
| 6 | 89 |

Figure 3.32 – Sorting a Series by value

Note the `sort_values` and `sort_index` methods also work for columns and indexes containing strings.

Using our `artist` DataFrame, let's try sorting a column with a string as follows:

```
sort_df = df.sort_values({by: "artist", inplace:false,
ascending: false})
table(sort_df)
```

The DataFrame is sorted using the `artist` column in descending order. This results in a table with `Rihanna` in the first row followed by `Passenger`, based on the first character:

| | artist | rating | dollar |
|---|---|---|---|
| 1 | Rihanna | 4.5 | $2m |
| 4 | Passenger | 5 | $3m |
| 2 | Gambino | 4 | $0.1m |
| 0 | Drake | 5 | $1m |
| 3 | Bellion | 5 | $2.5m |

Figure 3.33 – Sorting by the string column

In this section, we saw how to sort a DataFrame by column values and row index. In the next section, we will see how we can filter DataFrame values.

# Filtering

Filtering out rows of a DataFrame based on some particular values in a column comes in handy most of the time during data manipulation and processing. Danfo.js has a method called `query` that is used to filter rows based on the value in a column.

The `query` method takes in the following keyword arguments:

- `column`: The column name
- `is`: Specifies the logical operator to use (`>`, `<`, `>=`, `<=`, `==`)
- `to`: The value used to filter the DataFrame
- `inplace`: Updates the original DataFrame or returns a new one

Here is an example of how the `query` method works.

First, we create a DataFrame:

```
data = {"A": [30, 1, 2, 3],
        "B": [34, 4, 5, 6],
        "C": [20, 20, 30, 40]}

df = new dfd. DataFrame (data)
```

The following figure shows the table:

|   | A | B | C |
|---|---|---|---|
| 0 | 30 | 34 | 20 |
| 1 | 1 | 4 | 20 |
| 2 | 2 | 5 | 30 |
| 3 | 3 | 6 | 40 |

Figure 3.34 – DataFrame

Let's sort the DataFrame by a value in column B:

```
query_df = df.query({ column: "B", is: ">", to: 5 })
table(query_df)
```

Here, we filter column B by a value greater than 5. This will result in returning rows containing a value greater than 5 in column B, as shown here:

|   | A | B | C |
|---|---|---|---|
| 0 | 30 | 34 | 20 |
| 3 | 3 | 6 | 40 |

Figure 3.35 – Filtered by a value greater than 5 in column B

Let's filter the DataFrame by checking whether the values in column C equal to 20:

```
query_df = df.query({ column: "C", is: "==", to: 20})
table(query_df)
```

With this, we obtain the following output:

|   | A | B | C |
|---|---|---|---|
| 0 | 30 | 34 | 20 |
| 1 | 1 | 4 | 20 |

Figure 3.36 – Filtered by values equal to 20 in column C

In this section, we saw how to filter a DataFrame by column values. In the next section, we will see how to perform different arithmetic operations.

# Arithmetic operations

In this sub-section, we will look into different arithmetic operations and how to use them to preprocess our data.

Arithmetic operations such as addition, subtraction, multiplication, and division can be done between the following:

- A DataFrame and a Series

- A DataFrame and an array

- A DataFrame and a scalar value

We start by performing an arithmetic operation between DataFrame and scalar values:

```
data = {"Col1": [10, 45, 56, 10],
        "Col2": [23, 20, 10, 24]}
ar_df = new dfd. DataFrame (data)
//add a scalar variable
add_df = ar_df.add(20)
table(add_df)
```

The add method is used to add a scalar variable of 20 across each row of the DataFrame. The add method takes in two arguments:

- other: This represents either a DataFrame, Series, array, or scalar.

- axis: Specifies whether the operation should be applied by row or by column. The row axis is represented with 0 and the column by 1. The default is 0.

For the scalar operation, we obtain the following result:

|   | Col1 | Col2 |
|---|------|------|
| 0 | 10 | 23 |
| 1 | 45 | 20 |
| 2 | 56 | 10 |
| 3 | 10 | 24 |

|   | Col1 | Col2 |
|---|------|------|
| 0 | 30 | 43 |
| 1 | 65 | 40 |
| 2 | 76 | 30 |
| 3 | 30 | 44 |

Figure 3.37 – Left: original DataFrame. Right: DataFrame after doing scalar addition

Let's create a Series and pass it to the add method:

```
add_series = new dfd.Series([20,30])
add_df = ar_df.add(add_series, axis=1)
table(add_df)
```

A Series is created containing two-row elements. This two-row element equals the number of columns in the ar_df DataFrame, which means that the first row value of add_series belongs to the first column of ar_df, just as the second row value corresponds to the second column of ar_df.

The explanation in the preceding paragraph means that 20 will be used to multiply Col1 and 30 will be used to multiply Col2. For this to occur, we specify the axis to be 1, which tells the add operation to perform the operation column-wise, as we can see in the following table:

|   | Col1 | Col2 |
|---|------|------|
| 0 | 30   | 53   |
| 1 | 65   | 50   |
| 2 | 76   | 40   |
| 3 | 30   | 54   |

Figure 3.38 – Add operation between the DataFrame and Series

The addition operation between a DataFrame and a Series is the same between a DataFrame and a normal JavaScript array, since a Series is just a JavaScript array with some extended functionality.

Let's look into the operation between two DataFrames:

```
data = {"Col1": [1, 4, 5, 0],
        "Col2": [2, 0, 1, 4]}

data2 = {"new_col1": [1, 5, 20, 10],
         "new_Col2": [20, 2, 1, 2]}
df = new dfd.DataFrame(data)
df2 = new dfd.DataFrame(data2)

df_new = df.add(df2)
```

First, we create two sets of DataFrames and then add them together; but we must ensure the two DataFrames have the same number of columns and rows. Also, this operation is done row-wise.

The operation does not actually care whether the two DataFrames have the same column name or not, but the resulting column name is the column of the DataFrame we are adding to the former DataFrame.

The following table shows the result of the DataFrames created in the preceding code:

| | Col1 | Col2 | | new_col1 | new_Col2 | | Col1 | Col2 |
|---|---|---|---|---|---|---|---|---|
| 0 | 1 | 2 | 0 | 1 | 20 | 0 | 2 | 22 |
| 1 | 4 | 0 | 1 | 5 | 2 | 1 | 9 | 2 |
| 2 | 5 | 1 | 2 | 20 | 1 | 2 | 25 | 2 |
| 3 | 0 | 4 | 3 | 10 | 2 | 3 | 10 | 6 |

Figure 3.39 – From left to right: tables for DataFrames df and df2 and addition between df and df2

The example done with the add operation is the same for the sub(), mul(), div(), pow(), and mod() methods.

Also, for both DataFrames and Series there is a set of mathematical operations called cumulative, which comprises the following methods:

- Cumulative sum: cumsum()
- Cumulative min: cummin()
- Cumulative max: cummax()
- Cumulative product: cumprod()

Each of these operations is the same in terms of the argument being passed in. They all accept the axis in which the operation is to be performed:

```
data = [[11, 20, 3], [1, 15, 6], [2, 30, 40], [2, 89, 78]]
cols = ["A", "B", "C"]
df = new dfd. DataFrame (data, { columns: cols })
new_df = df.cumsum({ axis: 0 })
```

We create a DataFrame by specifying its data and columns, and we then perform a cumulative sum along the 0 axis.

The following table shows the result of the preceding code:

| | A | B | C |
|---|---|---|---|
| 0 | 11 | 20 | 3 |
| 1 | 1 | 15 | 6 |
| 2 | 2 | 30 | 40 |
| 3 | 2 | 89 | 78 |

| | A | B | C |
|---|---|---|---|
| 0 | 11 | 20 | 3 |
| 1 | 12 | 35 | 9 |
| 2 | 14 | 65 | 49 |
| 3 | 16 | 154 | 127 |

Figure 3.40 – Left: the DataFrame before cumsum(). Right: the DataFrame after cumsum()

Let's perform a cumsum operation along axis 1 of the DataFrame:

```
new_df = df.cumsum({ axis: 1 })
```

This gives us the following table:

| | A | B | C |
|---|---|---|---|
| 0 | 11 | 31 | 34 |
| 1 | 1 | 16 | 22 |
| 2 | 2 | 32 | 72 |
| 3 | 2 | 91 | 169 |

Figure 3.41 – cumsum along axis 1

The same cumulative operations can also be applied to Series. Let's apply the same cumulative sum to Series data:

```
series = new dfd.Series([2,3,4,5,6,7,8,9])
c_series = series.cumsum()
table(c_series)
```

cumsum in a Series does not take in any argument. With the preceding operation, we have as the output, the following table *Figure 3.42* that is the original Series before applying the cumulative Series.

The following table shows the result of the `series.cumsum()` operation in the preceding code:

| | 0 |
|---|---|
| 0 | 2 |
| 1 | 5 |
| 2 | 9 |
| 3 | 14 |
| 4 | 20 |
| 5 | 27 |
| 6 | 35 |
| 7 | 44 |

Figure 3.42 – The cumulative sum Series

All other cumulative operations, such as `cumprod`, `cummin`, and `cummax`, are used in the same way as the `cumsum()` operation as shown in the preceding `cumsum` examples.

In this section, we looked into different arithmetic operations and how to make use of these operations to preprocess data. In the next section, we will dive into logical operations such as how to perform logical operations between a DataFrame and Series, array, and `Scalar` values.

## Logical operations

In this section, we will see how we can perform logical operations, such as comparing the values between a DataFrame and Series, array, and scalar values.

The way logical operations are called and used is quite similar to how arithmetic operations work. A logical operation can be done between the following:

- A DataFrame and a Series
- A DataFrame and an array
- A DataFrame and a scalar value

The following logical operations are implemented:

- Equal to (==): `.eq()`
- Greater than (>): `gt()`
- Less than (<): `lt()`

- Not equal to (!=): ne ()

- Less than or equal to (<=): le ()

- Greater than or equal to (>=): ge ()

All these methods take in the same argument, which are other and axis. Using the lt () method, let's see how they work:

```
data = {"Col1": [10, 45, 56, 10],
        "Col2": [23, 20, 10, 24]}
df = new dfd. DataFrame (data)
df_rep = df.lt(20)
table(df_rep)
```

The code checks whether each of the values in the DataFrame is less than 20. The result can be seen in the following figure:

|   | Col1 | Col2 |   | Col1 | Col2 |
|---|------|------|---|------|------|
| 0 | 10   | 23   | 0 | true | false |
| 1 | 45   | 20   | 1 | false | false |
| 2 | 56   | 10   | 2 | false | true |
| 3 | 10   | 24   | 3 | true | false |

Figure 3.43 – Left: DataFrame for df. Right: DataFrame for df_rep

Also, the same operation can be done with a DataFrame and Series, as follows:

```
series = new dfd.Series([45,10])
df_rep = df.lt(series, axis=1)
table(df_rep)
```

The series operation with the DataFrame is done column-wise. The following table is the result:

|   | Col1 | Col2 |
|---|------|------|
| 0 | true | false |
| 1 | false | false |
| 2 | false | false |
| 3 | true | false |

Figure 3.44 – Logical operation between a DataFrame and Series

Let's perform a logical operation between two DataFrames, as follows:

```
data = {"Col1": [10, 45, 56, 10],
        "Col2": [23, 20, 10, 24]}
data2 = {"new_col1": [10, 45, 200, 10],
         "new_Col2": [230, 200, 110, 24]}

df = new dfd.DataFrame (data)
df2 = new dfd.DataFrame (data2)

df_rep = df.lt(df2)
table(df_rep)
```

The code outputs the following result:

| | Col1 | Col2 |
|---|---|---|
| 0 | false | true |
| 1 | false | true |
| 2 | true | true |
| 3 | false | false |

Figure 3.45 – Logical operation between DataFrames

In this section, we looked into some essential DataFrame operations, such as how to filter a DataFrame by column value. We also saw how to sort a DataFrame by column value and row index. Also, this section dived into the indexing of DataFrames with `loc` and `iloc`, and finally, we looked into arithmetic and logical operations.

In the next section, we will dive into working with different data formats, how to parse these files into Danfo.js, and converting them into a DataFrame or Series.

# Data loading and working with different file formats

In this section, we will look at how to work with different file formats. Danfo.js provides a method for working with three different file formats: CSV, Excel, and JSON. Data presented in these formats can easily be read and presented as a DataFrame.

The following are the methods needed for each of the file formats:

- `read_csv()`: To read CSV files
- `read_json()`: To read JSON files
- `read_excel()`: To read Excel files presented as `.xslx`

Each of these methods can read data locally and from the internet. Also, these methods only have access to the local files in a `Node.js` environment. On the web, these methods can read provided files (CSV, JSON, and .xslx files) provided these files are available on the internet.

Let's see how to read local files in the Node.js environment:

```
const dfd = require("Danfo.Js-node")
dfd.read_csv('titanic.csv').then(df => {
  df.head().print()
})
```

First, we import the Node.js version of Danfo.js and then make a call to `read_csv()` to read the `titanic.csv` files available in the same directory. Note that if the file to be read is not in the same directory, you will need to specify the path.

`df.head().print()` prints the first five rows of the DataFrame table in the Node.js console; the `print()` function is similar to the `table()` function we've been using in Dnotebook. We obtain the following table from the preceding code:

| | Survived | Pclass | Name | Sex | ... |
|---|---|---|---|---|---|
| 0 | 0 | 3 | Mr. Owen Harr... | male | ... |
| 1 | 1 | 1 | Mrs. John Bra... | female | ... |
| 2 | 1 | 3 | Miss. Laina H... | female | ... |
| 3 | 1 | 1 | Mrs. Jacques ... | female | ... |
| 4 | 0 | 3 | Mr. William H... | male | ... |

Figure 3.46 – DataFrame table read from a CSV file

Similarly, the same thing can be done on the web, but we will be using the `http` link, which makes the same data available online:

```
csvUrl =
      "https://storage.googleapis.com/tfjs-examples/
multivariate-linear-regression/data/boston-housing-train.csv";
dfd.read_csv(csvUrl).then((df) => {
   df.print()
});
```

This gives the same result as the previous code block. The same thing applies to all other file format methods:

```
const dfd = require("Danfo.Js-node")

dfd.read_excel('SampleData.xlsx', {header_index: 7}).then(df =>
{
   df.head().print()
})
```

The `read_excel()` method accepts an optional config argument that ensures the Excel files are properly parsed:

- `source`: String, URL, or local file path to retrieve an Excel file.
- `sheet_name` (optional): The name of the sheet you want to parse. The default is the first sheet.
- `header_index` (optional): int, index of the row that represents the header column.
- `data_index` (optional): int, index of the row that indicates where the data begins.

In the preceding code block, we specify the `header_index` value to be 7 since that's where the header columns are located, hence we have the following result:

| | Year | Stocks | T.Bills | T.Bonds |
|---|---|---|---|---|
| 0 | 1928 | 0.4381 | 0.0308 | 0.0084 |
| 1 | 1929 | −0.083 | 0.0316 | 0.042 |
| 2 | 1930 | −0.2512 | 0.0455 | 0.0454 |
| 3 | 1931 | −0.4384 | 0.0231 | −0.0256 |
| 4 | 1932 | −0.0864 | 0.0107 | 0.0879 |

Figure 3.47 – DataFrame table read from an Excel file

The `read_json()` method is quite similar to `read_csv()` in terms of the argument being passed in; the method only accepts a URL to the JSON file or a directory path to where the JSON file is located:

```
const dfd = require("Danfo.Js-node")
dfd.read_json('book.json').then(df => {
  df.head().print()
})
```

This reads a file called `book.json` from the same directory and outputs the following:

| | book_id | title | image_url | authors |
|---|---|---|---|---|
| 0 | 1 | Harry Potter ... | https://image... | J.K. Rowling,... |
| 1 | 2 | Harry Potter ... | https://image... | J.K. Rowling,... |
| 2 | 3 | Harry Potter ... | https://image... | J.K. Rowling,... |
| 3 | 5 | Harry Potter ... | https://image... | J.K. Rowling,... |
| 4 | 6 | Harry Potter ... | https://image... | J.K. Rowling,... |

Figure 3.48 – DataFrame read from a JSON file

Also, Danfo.js contains an ultimate tabular file reader method called `reader`; this method can read both CSV and Excel. It can also read other files such as `Datapackage` as specified in the Frictionless specs at `https://frictionlessdata.io/`.

The `reader` method makes use of a package called `Frictionless.js` at `https://github.com/frictionlessdata/frictionless-js` to read local or remote files.

The `reader` method has the same API design as `read_csv` and `read_excel`, just that it can read both file types, as shown in the following code:

```
const dfd = require("Danfo.Js-node")
// for reading csv
dfd.read('titanic.csv').then(df => {
   df.head().print()
})
```

From the preceding code, we output the same table as shown in *Figure 3.48*.

# Transforming a DataFrame into another file format

After a series of data processing, we might want to transform our final DataFrame into a file format for proper saving. Danfo.js implements a method to transform a DataFrame into CSV format.

Let's convert the DataFrame we've created so far into a CSV file. Note that this will only work in a Node.js environment:

```
df.to_csv("/home/link/to/path.csv").then((csv) => {
     console.log(csv);
}).catch((err) => {
     console.log(err);
})
```

This (`df.to_csv()`) saves the DataFrame in a CSV file in the path directory specified and with the name given to it (`path.csv`).

In this section, we looked into reading files in different formats, such as CSV, JSON, and Excel, as available in Danfo.js. We looked into different methods of reading these files and we also looked into a more generic method for reading these file formats.

We also saw how to convert a DataFrame into the file format specified by the user.

# Summary

In this chapter, we went through why Danfo.js is needed, then looked into what Series and DataFrames actually are. We also discussed some of the essential functionality available in Danfo.js and implemented it in DataFrames and Series.

We also saw how we can use DataFrames and Series methods to handle and preprocess data. We saw how to filter a DataFrame based on column values. We also sorted a DataFrame by row index and column values. This chapter equips us to perform day-to-day data operations such as reading files in different formats, converting formats, and saving DataFrames after preprocessing into a desirable file format.

In the next chapter, we will look into data analysis, wrangling, and transformation. We will discuss data handling and preprocessing further and see how to handle missing numbers and how to deal with string and time series data.

# 4
# Data Analysis, Wrangling, and Transformation

**Data analysis**, **wrangling**, and **transformation** are important aspects of any data-driven project, and the majority of your time as a data analyst/scientist will be spent doing one form of data processing or the other. While JavaScript is a flexible language with good features for manipulating data structures, it is quite tedious to write utility functions to perform data wrangling operations all the time. As such, we have built powerful data wrangling and transformation features into Danfo.js, and this can greatly reduce the time that's spent on this stage.

In this chapter, we will show you how to practically use Danfo.js on real-world datasets. You'll learn how to load different types of datasets and analyze them by performing operations such as handling missing values, calculating descriptive statistics, performing mathematical operations, combining datasets, and performing string manipulations.

In this chapter, we will cover the following topics:

- Transforming data
- Combining datasets
- Series data accessors
- Calculating statistics

# Technical requirements

To follow along with this chapter, you should have the following:

- A modern browser such as Chrome, Safari, Opera, or Firefox
- Node.js, Danfo.js, and Dnotebook installed on your system
- A stable internet connection for downloading datasets

The installation instructions for Danfo.js can be found in *Chapter 3*, *Getting Started with Danfo.js*, while the installation steps for Dnotebook can be found in *Chapter 2*, *Dnotebook – An Interactive Computing Environment for JavaScript*.

> **Note**
>
> If you do not want to install any software or library, you can use the online version of Dnotebook at `https://playnotebook.jsdata.org/`. However, don't forget to install the latest version of Danfo.js before using any functionality!

# Transforming data

Data transformation is the process of converting data from one format (master format) into another (target format) based on defined steps/processes. Data transformation can be simple or complex, depending on the structure, format, end goal, size, or complexity of the dataset, and as such, it is important to know the features that are available in Danfo.js for doing these transformations.

In this section, we'll introduce some features available in Danfo.js for doing data transformation. Under each sub-section, we'll introduce a couple of functions, including `fillna`, `drop_duplicates`, `map`, `addColumns`, `apply`, `query`, and `sample`, as well as functions for encoding data.

# Replacing missing values

Many datasets come with missing values and in order to get the most out of these datasets, we must do some form of data filling/replacement. Danfo.js provides a `fillna` method that, when given a DataFrame or Series, can automatically fill any missing field with the specified value.

When you load a dataset into Danfo.js data structures, all the missing values, which can be undefined, empty, null, none, and so on, are stored as NaN, and as such, the `fillna` method can easily find and replace them.

## Replacing values in a Series

The signature of the `fillna` method of a `Series` object is as follows:

```
Series.fillna({value, inplace})
```

The `value` parameter is the new value you want to use for replacing missing values, while the `inplace` parameter is used to specify whether to make changes directly to the object or a copy. Let's look at an example of this.

Let's assume we have the following Series:

```
sdata = new dfd.Series([NaN, 1, 2, 33, 4, NaN, 5, 6, 7, 8])
sdata.print() //use print to show series in browser or node environment
table(sdata) //use table to show series in Dnotebook environment
```

In your Dnotebook environment, `table(sdata)` will display the table shown in the following figure:

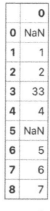

|   | 0 |
|---|---|
| 0 | NaN |
| 1 | 1 |
| 2 | 2 |
| 3 | 33 |
| 4 | 4 |
| 5 | NaN |
| 6 | 5 |
| 7 | 6 |
| 8 | 7 |

Figure 4.1 – Series with missing values

> **Note**
> If you are working in a browser or the Node.js environment, you can view the output of the `print` function in the console.

To replace the missing values (NaN), we can do the following:

```
sdata_new = sdata.fillna({ value: -999})
table(sdata_new)
```

Printing the output gives us the following table:

|   | 0 |
|---|---|
| 0 | -999 |
| 1 | 1 |
| 2 | 2 |
| 3 | 33 |
| 4 | 4 |
| 5 | -999 |
| 6 | 5 |
| 7 | 6 |
| 8 | 7 |

Figure 4.2 – Series with missing values

You can also fill missing values in place, that is, you can directly mutate the object, instead of creating a new one. This can be seen in the following code:

```
sdata.fillna({ value: -999, inplace: true})
table(sdata)
```

Filling in place can help reduce memory usage in scenarios where you are working with large DataFrames or Series.

## Replacing values in DataFrames

The `fillna` function can also be used to fill in missing values in a DataFrame. The signature of the `fillna` method of a DataFrame object is as follows:

```
DataFrame.fillna({columns, value, inplace})
```

First, let's understand the parameters:

- `columns`: The `columns` parameter is the array of column names that you want to fill.

- `values`: The `values` parameter, which is also an array and must be the same size as `columns`, holds the corresponding values you want to replace with.

- `inplace`: This specifies whether we should modify the current DataFrame or return a new one.

Now, let's look at an example.

Let's assume we have the following DataFrame:

```
data = {"Name":["Apples", "Mango", "Banana", undefined],
        "Count": [NaN, 5, NaN, 10],
        "Price": [200, 300, 40, 250]}

df = new dfd.DataFrame(data)
table(df)
```

The output of the code is shown in the following figure:

|   | Name   | Count | Price |
|---|--------|-------|-------|
| 0 | Apples | NaN   | 200   |
| 1 | Mango  | 5     | 300   |
| 2 | Banana | NaN   | 40    |
| 3 | NaN    | 10    | 250   |

Figure 4.3 – DataFrame with missing values

In terms of filling in the missing values, there are two ways we can approach this. First, we can fill in all the missing fields with single values, as shown in the following code snippet:

```
df_filled = df.fillna({ values: [-99]})
table(df_filled)
```

The code output is shown in the following figure:

| | Name | Count | Price |
|---|---|---|---|
| 0 | Apples | -99 | 200 |
| 1 | Mango | 5 | 300 |
| 2 | Banana | -99 | 40 |
| 3 | -99 | 10 | 250 |

Figure 4.4 – Filling in all the missing values in a DataFrame with a single value

The majority of the time, when working with DataFrames, filling in all the missing values with the same field is not advisable or even useful, as you must take into consideration the fact that different fields have different types of values, which means the filling strategies will differ.

To handle such cases, we can specify a list of columns and their corresponding values to fill them, as we will show here.

Using the same DataFrame we used in the previous example, we can specify the Name and Count columns and fill them with the Apples and -99 values, as follows:

```
data = {"Name":["Apples", "Mango", "Banana", undefined],
        "Count": [NaN, 5, NaN, 10],
        "Price": [200, 300, 40, 250]}

df = new dfd.DataFrame(data)
df_filled = df.fillna({columns: ["Name", "Count"], values:
["Apples", -99]})
table(df_filled)
```

The code's output is shown in the following figure:

| | Name | Count | Price |
|---|---|---|---|
| 0 | Apples | -99 | 200 |
| 1 | Mango | 5 | 300 |
| 2 | Banana | -99 | 40 |
| 3 | Apples | 10 | 250 |

Figure 4.5 – Filling in the missing values in a DataFrame with specific values

> **Note**
> Fill values such as -9, -99, and -999 are commonly used to indicate missingness in data analytics.

Let's see how to remove duplicates in the next section.

## Removing duplicates

Duplicate fields are common scenarios when working with Series or columns in a DataFrame. For example, take a look at the Series in the following code:

```
data = [10, 45, 56, 10, 23, 20, 10, 10]
sf = new dfd.Series(data)
table(sf)
```

The code's output is shown in the following figure:

|   | **0** |
|---|---|
| **0** | 10 |
| **1** | 45 |
| **2** | 56 |
| **3** | 10 |
| **4** | 23 |
| **5** | 20 |
| **6** | 10 |
| **7** | 10 |

Figure 4.6 – Series with duplicate fields

In the preceding figure, you can see that the value 10 occurs multiple times. If the need arises, you can easily drop/remove these duplicates with the drop_duplicates function. The signature of the drop_duplicates function is as follows:

```
Series.drop_duplicates({inplace, keep)
```

The drop_duplicates function has just two parameters, and the first parameter (inplace) is pretty self-explanatory. The second parameter, keep, can be used to specify which of the duplicates to keep. You can either keep the first or last duplicate in a Series. This helps preserve the structure and order or values in a Series after dropping duplicates.

Let's see this in action. In the following code block, we are dropping all the duplicate values and keeping just the first occurrence:

```
data1 = [10, 45, 56, 10, 23, 20, 10, 10, 20, 20]
sf = new dfd.Series(data1)
sf_drop = sf.drop_duplicates({keep: "first"})
table(sf_drop)
```

The code's output is shown in the following figure:

|   | 0  |
|---|----|
| 0 | 10 |
| 1 | 45 |
| 2 | 56 |
| 4 | 23 |
| 5 | 20 |

Figure 4.7 – Series after dropping duplicates with keep set to first

Looking at the preceding output, you can see that the first value of the duplicates is kept while the others are dropped. In contrast, let's set the keep parameter to last and watch the order of the fields change:

```
sf_drop = sf.drop_duplicates({keep: "last"})
table(sf_drop)
```

The code's output is shown in the following figure:

|   | 0  |
|---|----|
| 1 | 45 |
| 2 | 56 |
| 4 | 23 |
| 7 | 10 |
| 9 | 20 |

Figure 4.8 – Series after dropping duplicates with keep set to last

Notice that the last 10 and 20 duplicates fields are kept, and that the order is different from when we set keep to first. The following figure will help you understand the keep parameter when dropping duplicates:

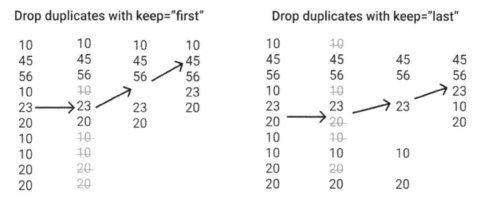

Figure 4.9 – Difference between the output when keep is set to first or last

From the preceding figure, you'll observe that the major difference between setting keep to first or last is the order of the resulting values.

In the next sub-section, we'll look at data transformation with the map function.

## Data transformation with the map function

In some cases, you may want to apply a transformation to each value in a Series or DataFrame column. This is usually useful when you have some custom function or mapping and want to apply it easily to each field. Danfo.js provides a simple interface called map that can be used here. Let's see an example.

Let's assume we have a DataFrame of items and their corresponding weights in grams, as shown in the following code:

```
df = new dfd.DataFrame({'item': ['salt', 'sugar', 'rice',
'apple', 'corn', 'bread'],
'grams': [400, 200, 120, 300, 70.5, 250]})
table(df)
```

The code's output is shown in the following figure:

| | item | grams |
|---|---|---|
| 0 | salt | 400 |
| 1 | sugar | 200 |
| 2 | rice | 120 |
| 3 | apple | 300 |
| 4 | corn | 70.5 |
| 5 | bread | 250 |

Figure 4.10 – DataFrame with items and their corresponding weights in grams

We want to create a new column called kilograms, whose values are the corresponding grams but converted into kilograms. Here, we can do the following:

1.  Create a convertToKg function with the following code:

    ```
    function convertToKg (gram) {
      return gram / 1000
    }
    ```

2.  Call the map function on the grams column, as shown in the following code block:

    ```
    kilograms = df ['grams'].map (convertToKg)
    ```

3.  Add the new kilograms column to the DataFrame using the addColumn function, as shown in the following code:

    ```
    df.addColumn({ "column": "kilograms", "value":
    kilograms });
    ```

    Putting this all together, we have the following code:

    ```
    df = new dfd.DataFrame({'item': ['salt', 'sugar', 'rice',
    'apple', 'corn', 'bread'],
    'grams': [400, 200, 120, 300, 70.5, 250]})

    function convertToKg (gram) {
      return gram / 1000
    }
    kilograms  = df ['grams'].map (convertToKg)
    df.addColumn({ "column": "kilograms", "value":
    ```

```
    kilograms    });
  table(df)
```

The code output is shown in the following figure:

| | item | grams | kilograms |
|---|---|---|---|
| 0 | salt | 400 | 0.4 |
| 1 | sugar | 200 | 0.2 |
| 2 | rice | 120 | 0.12 |
| 3 | apple | 300 | 0.3 |
| 4 | corn | 70.5 | 0.0705 |
| 5 | bread | 250 | 0.25 |

Figure 4.11 – DataFrame with items and their corresponding weights in grams

The map function can also perform one-to-one mapping when you pass it an object with key-value pairs. This can be used for things such as encoding, quick mapping, and so on. Let's see an example of this.

Let's assume we have the following Series:

```
sf = new dfd.Series([1, 2, 3, 4, 4, 4])
```

Here, we want to map each numeric value to its corresponding name. We can specify a mapper object and pass it to the map function, as shown in the following code:

```
mapper = { 1: "one", 2: "two", 3: "three", 4: "four" }
sf = sf.map(mapper)
table(sf)
```

The code's output is shown in the following figure:

| | 0 |
|---|---|
| 0 | one |
| 1 | two |
| 2 | three |
| 3 | four |
| 4 | four |
| 5 | four |

Figure 4.12 – Series with numeric values mapped to string names

The map function is very useful when working with Series, but sometimes, you may want to apply functions to specific axes (rows or columns). In such scenarios, you can use another Danfo.js function called the apply function.

In the next section, we'll introduce the apply function.

# Data transformation with the apply function

The apply function can be used to map functions or transformations to a specific axis in a DataFrame. It is a little more advanced and powerful than the map function, and we'll explain why.

First is the fact that we can apply tensor operations across a specified axis. Let's take a look at a DataFrame that contains the following values:

```
data = [[1, 2, 3], [4, 5, 6], [20, 30, 40], [39, 89, 78]]
cols = ["A", "B", "C"]
df = new dfd.DataFrame(data, { columns: cols })
table(df)
```

The code's output is shown in the following figure:

|   | A  | B  | C  |
|---|----|----|----|
| 0 | 1  | 2  | 3  |
| 1 | 4  | 5  | 6  |
| 2 | 20 | 30 | 40 |
| 3 | 39 | 89 | 78 |

Figure 4.13 – Sample DataFrame with three columns

Now, we can apply any compatible tensor operation across a specified axis. For example, in the following code, we can apply a softmax function (https://en.wikipedia.org/wiki/Softmax_function) to each element in the DataFrame:

```
function sum_vals(x) {
    return x.softmax()
}
let df_new = df.apply({axis: 0, callable: sum_vals })
table(df_new)
```

The code's output is shown in the following figure:

| | A | B | C |
|---|---|---|---|
| **0** | 0.09003057330846786 | 0.2447284758090973 | 0.6652409434318542 |
| **1** | 0.09003057330846786 | 0.2447284758090973 | 0.6652409434318542 |
| **2** | 2.0610599893444714e-9 | 0.00004539786823443137 | 0.9999545812606812 |
| **3** | 1.9287177078288054e-22 | 0.9999833106994629 | 0.000016701422282494605 |

Figure 4.14 – DataFrame after applying a softmax function element-wise

> **Note**
>
> We set the axis to 0 for element-wise tensor operations. This is because some tensor operations cannot be performed element-wise. You can read more about the supported operations at `https://js.tensorflow.org/api/latest/`.

Let's look at another example of applying a function that can work on both columns (axis = 1) and rows (axis = 0), respectively.

We'll use the TensorFlow sum function for this, as shown in the following code. First, let's apply it across the row axis:

```
data = [[1, 2, 3], [4, 5, 6], [20, 30, 40], [39, 89, 78]]
cols = ["A", "B", "C"]
df = new dfd.DataFrame(data, { columns: cols })


function sum_vals(x) {
    return x.sum()
}
df_new = df.apply({axis: 0, callable: sum_vals })
table(df_new)
```

Printing the df_new DataFrame results in the following output:

| | 0 |
|---|---|
| **0** | 6 |
| **1** | 15 |
| **2** | 90 |
| **3** | 206 |

Figure 4.15 – DataFrame after applying the sum function across the row (0) axis

Using the same DataFrame as earlier, change the axis to `1` for the column-wise operation, as shown in the following code snippet:

```
df_new = df.apply({axis: 1, callable: sum_vals })
table(df_new)
```

Printing the DataFrame results in the following output:

|   | 0   |
|---|-----|
| A | 64  |
| B | 126 |
| C | 127 |

Figure 4.16 – DataFrame after applying the sum function across the column (1) axis

Custom JavaScript functions can also be used with the `apply` function. The only caveat here is that you do not need to specify an axis, as JavaScript functions are applied element-wise.

Let's look at an example of using `apply` with a JavaScript function. In the following code block, we are applying the `toLowerCase` string function to each element in the DataFrame:

```
data = [{ short_name: ["NG", "GH", "EGY", "SA"] },
           { long_name: ["Nigeria", "Ghana", "Eqypt", "South
Africa"] }]
df = new dfd.DataFrame(data)
function lower(x) {
    return '${x}'.toLowerCase()
}
df_new = df.apply({ callable: lower })
table(df_new)
```

Printing the DataFrame results in the following output:

|   | short_name                    |
|---|-------------------------------|
| 0 | ng,gh,egy,sa                  |
| 1 | nigeria,ghana,eqypt,south africa |

Figure 4.17 – DataFrame after applying a JavaScript function element-wise

The `apply` function is very powerful and can be used to apply custom transformations to columns or rows across a DataFrame or Series. In the next sub-section, we'll look at the different ways of filtering and querying DataFrames and Series.

# Filtering and querying

Filtering and querying are important when we need to get a subset of the data that satisfies a specific Boolean condition. Filtering and querying can be done on DataFrames and Series using the `query` method, as we'll demonstrate in the following examples.

Let's assume we have a DataFrame that contains the following columns:

```
data = {"A": [30, 1, 2, 3],
        "B": [34, 4, 5, 6],
        "C": [20, 20, 30, 40]}
cols = ["A", "B", "C"]
df = new dfd.DataFrame(data, { columns: cols })
table(df)
```

Printing the DataFrame results in the following output:

|   | A  | B  | C  |
|---|----|----|----|
| 0 | 30 | 34 | 20 |
| 1 | 1  | 4  | 20 |
| 2 | 2  | 5  | 30 |
| 3 | 3  | 6  | 40 |

Figure 4.18 – DataFrame with sample values for querying

Now, let's filter the DataFrame and only return rows where the B column has values greater than 5. This should return rows 0 and 3. We can do this with the `query` function, as shown in the following code:

```
df_filtered = df.query({ column: "B", is: ">", to: 5})
table(df_filtered)
```

This gives us the following output:

| | A | B | C |
|---|---|---|---|
| **0** | 30 | 34 | 20 |
| **3** | 3 | 6 | 40 |

Figure 4.19 – DataFrame after querying

The `query` method accepts all JavaScript Boolean operators (>, <, >=, <=, and ==), and it also works on string columns, as shown in the following example.

First, let's create a DataFrame with string columns:

```
data = {"A": ["Ng", "Yu", "Mo", "Ng"],
        "B": [34, 4, 5, 6],
        "C": [20, 20, 30, 40]}


df = new dfd.DataFrame(data)
table(df)
```

The output is shown in the following figure:

| | A | B | C |
|---|---|---|---|
| **0** | Ng | 34 | 20 |
| **1** | Yu | 4 | 20 |
| **2** | Mo | 5 | 30 |
| **3** | Ng | 6 | 40 |

Figure 4.20 – DataFrame before applying the string query

Next, we will run a query with the equals to operator (`"=="`):

```
query_df = df.query({ column: "A", is: "==", to: "Ng"})
table(query_df)
```

This results in the following output:

| | A | B | C |
|---|---|---|---|
| 0 | Ng | 34 | 20 |
| 3 | Ng | 6 | 40 |

Figure 4.21 – DataFrame after applying the string query

In most scenarios, the DataFrame you're querying is large, so you may want to perform a query in place. This can be done by specifying `inplace` as `true`, as shown in the following code block:

```
data = {"A": [30, 1, 2, 3],
        "B": [34, 4, 5, 6],
        "C": [20, 20, 30, 40]}

cols = ["A", "B", "C"]
df = new dfd.DataFrame(data, { columns: cols })
df.query({ column: "B", is: "==", to: 5, inplace: true })
table(df)
```

Printing the DataFrame results in the following output:

| | A | B | C |
|---|---|---|---|
| 2 | 2 | 5 | 30 |

Figure 4.22 – DataFrame after performing an inplace query

The `query` method is very important, and you use it a lot when filtering your data by specific properties. Another bonus of using the `query` function is that it allows `inplace` functionality, which means it's useful for filtering large datasets.

In the next sub-section, we'll look at another useful concept called random sampling.

# Random sampling

**Random sampling** from a DataFrame or Series is useful when you need to reorder rows randomly. This is mostly useful in preprocessing steps before **machine learning (ML)**.

Let's see an example of using random sampling. Let's assume we have a DataFrame that contains the following values:

```
data = [[1, 2, 3], [4, 5, 6], [20, 30, 40], [39, 89, 78]]
```

```
cols = ["A", "B", "C"]
df = new dfd.DataFrame(data, { columns: cols })
table(df)
```

This results in the following output:

| | A | B | C |
|---|---|---|---|
| 0 | 1 | 2 | 3 |
| 1 | 4 | 5 | 6 |
| 2 | 20 | 30 | 40 |
| 3 | 39 | 89 | 78 |

Figure 4.23 – DataFrame before random sampling

Now, let's randomly select two rows by calling the `sample` function on the DataFrame, as shown in the following code:

```
async function load_data() {
  let data = {
    Name: ["Apples", "Mango", "Banana", "Pear"],
    Count: [21, 5, 30, 10],
    Price: [200, 300, 40, 250],
  };

  let df = new dfd.DataFrame(data);
  let s_df = await df.sample(2);
  s_df.print();


}
load_data()
```

This results in the following output:

| | Name | Count | Price |
|---|---|---|---|
| 0 | Apples | 21 | 200 |
| 2 | Banana | 30 | 40 |

Figure 4.24 – DataFrame in browser console after randomly sampling two rows

In the preceding code, you'll notice that the code is wrapped in an `async` function. This is because the `sample` method returns a promise, and as such, we have to await the result. The preceding code can be executed in the browser or node.js environment as is, but it will need a little bit of tweaking to work in Dnotebook.

If you want to run the exact code in Dnotebook, you can tweak it like so:

```
var sample;
async function load_data() {
  let data = {
    Name: ["Apples", "Mango", "Banana", "Pear"],
    Count: [21, 5, 30, 10],
    Price: [200, 300, 40, 250],
  };


  let df = new dfd.DataFrame(data);
  let s_df = await df.sample(2);
  sample = s_df


}
load_data()
```

In the preceding code, you can see that we explicitly defined `sample` using `var`. This makes the `sample` variable available to all cells. Using the `let` declaration here instead will make the variable available to only the cell where it was defined.

Next, in a new cell, you can print the sample DataFrame using the `table` method, as shown in the following code:

```
table(sample)
```

This results in the following output:

|   | Name | Count | Price |
|---|------|-------|-------|
| 0 | Apples | 21 | 200 |
| 2 | Banana | 30 | 40 |

Figure 4.25 – DataFrame in Dnotebook after randomly sampling two rows

In the next section, we'll briefly talk about some encoding features available in Danfo.js.

# Encoding DataFrames and Series

**Encoding** is another important transformation that can be applied to DataFrames/Series before ML or **statistical modeling**. It is an important data preprocessing step that is always performed before feeding data into models.

ML or statistical models can only work with numeric values and as such, all string/ categorical columns must be converted appropriately into numeric form. There are numerous types of encoding, such as **one-hot encoding**, **label encoding**, **mean encoding**, and so on, and the choice of encoding may differ, depending on the type of data you have.

Danfo.js currently supports two popular encoding schemes: *label* and *one-hot encoder*, and in the following section, we'll explain how to use them.

## Label encoder

The label encoder maps categories in a column to an integer value between 0 and the number of unique classes in the column. Let's see an example of using this on a DataFrame.

Let's assume we have the following DataFrame:

```
data = { fruits: ['pear', 'mango', "pawpaw", "mango", "bean"] ,
         Count: [20, 30, 89, 12, 30],
         Country: ["NG", "NG", "GH", "RU", "RU"]}
df = new dfd.DataFrame(data)
table(df)
```

Printing the DataFrame results in the following output:

| | fruits | Count | Country |
|---|---|---|---|
| 0 | pear | 20 | NG |
| 1 | mango | 30 | NG |
| 2 | pawpaw | 89 | GH |
| 3 | mango | 12 | RU |
| 4 | bean | 30 | RU |

Figure 4.26 – DataFrame before encoding

Now, let's encode the `fruits` column using `LabelEncoder`, as shown in the following code block:

```
encode = new dfd.LabelEncoder()
encode.fit(df['fruits'])
```

In the preceding code block, you will notice that we created a `LabelEncoder` object first and then called the `fit` method on the column (`fruits`). The `fit` method simply learns and stores the mapping in the encoder object. This can be used later to transform any column/array, as we'll see shortly.

After calling the `fit` method, we must call the `transform` method to apply the labels, as shown in the following code block:

```
sf_enc = encode.transform(df['fruits'])
table(sf_enc)
```

Printing the DataFrame results in the following output:

|   | 0 |
|---|---|
| **0** | 0 |
| **1** | 1 |
| **2** | 2 |
| **3** | 1 |
| **4** | 3 |

Figure 4.27 – DataFrame before label encoding

From the preceding output, you can see that unique integers have been assigned to each label. In scenarios where `transform` is called on a column with a new category that was not available during the `fit` (learning) stage, it is represented with a value of $-1$. Let's take a look at an example of this while using the same encoder from the preceding example:

```
new_sf = encode.transform(["mango","man", "car", "bean"])
table(new_sf)
```

Printing the DataFrame results in the following output:

| | 0 |
|---|---|
| **0** | 1 |
| **1** | -1 |
| **2** | -1 |
| **3** | 3 |

Figure 4.28 – DataFrame after applying transform to an array with new categories

In the preceding example, you can see that we are calling the trained encoder on an array with new categories (man and car). Notice that the output has −1 in place of these new categories, as we explained earlier.

Next, let's talk about one-hot encoding.

## One-hot encoding

One-hot encoding is mostly applied to ordinal categories in a dataset. In this encoding scheme, binary values of 0 (hot) and 1 (cold) are used to encode unique categories.

Using the same DataFrame as in the preceding section, let's create a one-hot encoder object and apply it to the country column, as shown in the following code block:

```
encode = new dfd.OneHotEncoder()
encode.fit(df['country'])
sf_enc = encode.transform(df['country'])
table(sf_enc)
```

Printing the DataFrame results in the following output:

| | NG | GH | RU |
|---|---|---|---|
| **0** | 1 | 0 | 0 |
| **1** | 1 | 0 | 0 |
| **2** | 0 | 1 | 0 |
| **3** | 0 | 0 | 1 |
| **4** | 0 | 0 | 1 |

Figure 4.29 – DataFrame after one-hot encoding

From the preceding output, you can see that for each unique category, a series of 1s and 0s are used to replace the categories, and that we now have two extra columns that have been generated. The following diagram gives us an intuitive understanding of how each unique class is mapped to one-hot categories:

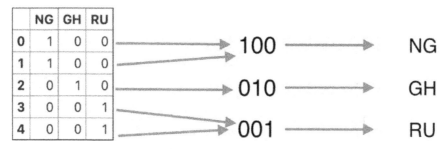

Figure 4.30 – One-hot encoding mapping for three categories

One last encoding feature available in Danfo.js that we'll talk about is the get_dummies function. This function works in the same way as the one-hot encoding function, but the major difference is the fact that you can apply it across a DataFrame and it will automatically encode any categorical column it finds.

Let's see an example of using the get_dummies function. Let's assume we have the following DataFrame:

```
data = { fruits: ['pear', 'mango', "pawpaw", "mango", "bean"] ,
            count: [20, 30, 89, 12, 30],
            country: ["NG", "NG", "GH", "RU", "RU"]}
df = new dfd.DataFrame(data)
table(df)
```

Printing the DataFrame results in the following output:

|   | fruits | count | country |
|---|--------|-------|---------|
| 0 | pear   | 20    | NG      |
| 1 | mango  | 30    | NG      |
| 2 | pawpaw | 89    | GH      |
| 3 | mango  | 12    | RU      |
| 4 | bean   | 30    | RU      |

Figure 4.31 – DataFrame before applying the get_dummies function

Now, we can encode all the categorical columns (`fruits` and `country`) at once by passing the DataFrame to the `get_dummies` function, as shown in the following code block:

```
df_enc = dfd.get_dummies({data: df})
table(df_enc)
```

Printing the DataFrame results in the following output:

| | count | fruits_pear | fruits_mango | fruits_pawpaw | fruits_bean | country_NG | country_GH | country_RU |
|---|---|---|---|---|---|---|---|---|
| 0 | 20 | 1 | 0 | 0 | 0 | 1 | 0 | 0 |
| 1 | 30 | 0 | 1 | 0 | 0 | 1 | 0 | 0 |
| 2 | 89 | 0 | 0 | 1 | 0 | 0 | 1 | 0 |
| 3 | 12 | 0 | 1 | 0 | 0 | 0 | 0 | 1 |
| 4 | 30 | 0 | 0 | 0 | 1 | 0 | 0 | 1 |

Figure 4.32 – One-hot encoding a mapping for three categories

From the preceding figure, you can see that the `get_dummies` function automatically detected the `fruits` and `country` categorical variables and one-hot encoded them. Columns are autogenerated and their names start with the corresponding column and category.

> **Note**
>
> You can specify the columns you want to encode, as well as the prefix for naming each encoded column. To see more available options, please refer to the official Danfo.js documentation: `https://danfo.jsdata.org/`.

In the next section, we'll look at various ways of combining datasets.

# Combining datasets

DataFrames and Series can be combined using built-in functions in Danfo.js. Methods such as `danfo.merge` and `danfo.concat` exist that, depending on the configurations, can help you combine datasets in different forms using familiar database-like joins.

In this section, we'll briefly talk about these join types, starting with the `merge` function.

# DataFrame merge

The `merge` operation is similar to the database `Join` operation in that it performs join operations on columns or indexes found in the object. The signature of the `merge` operation is as follows:

```
danfo.merge({left, right, on, how})
```

Let's understand what each parameter entails:

- `left`: The left-hand side DataFrame/Series you want to merge to.
- `right`: The right-hand side DataFrame/Series you want to merge to.
- `on`: The name(s) of the column(s) to join. These column(s) must be found in both the left and right DataFrames.
- `how`: The `how` parameter specifies how the merge should be carried out. It is similar to the database-style joins, and it can be either `left`, `right`, `outer`, or `inner`.

The Danfo.js `merge` function is very similar to the pandas `merge` function since it performs in-memory join operations that are similar to relational database joins. The following table provides a comparison of Danfo's merge methods and SQL joins:

| Danfo.js merge methods | SQL join name | Description |
|---|---|---|
| Inner | INNER JOIN | The intersection of keys from both DataFrames |
| Outer | FULL OUTER JOIN | The union of keys from both DataFrames |
| Right | RIGHT OUTER JOIN | Use keys from right frame only |
| Left | LEFT OUTER JOIN | Use keys from left frame only |

Figure 4.33 – Comparing the Danfo.js merge methods to SQL joins

(Source: pandas Doc: https://pandas.pydata.org/pandas-docs/stable/user_guide/merging.html#brief-primer-on-merge-methods-relational-algebra)

In the following sections, we'll present some examples to help you understand how and when to use the `merge` function.

## Inner merge by a single key

Merging DataFrames on a single common key results in an inner join by default. An inner join requires two DataFrames to have matching column values, and it returns the intersection of the two; that is, it returns a DataFrame with only those rows that have common characteristics.

Let's assume we have two DataFrames, as shown in the following code:

```
data = [['K0', 'k0', 'A0', 'B0'], ['k0', 'K1', 'A1', 'B1'],
         ['K1', 'K0', 'A2', 'B2'], ['K2', 'K2', 'A3', 'B3']]
data2 = [['K0', 'k0', 'C0', 'D0'], ['K1', 'K0', 'C1', 'D1'],
         ['K1', 'K0', 'C2', 'D2'], ['K2', 'K0', 'C3', 'D3']]
colum1 = ['Key1', 'Key2', 'A', 'B']
colum2 = ['Key1', 'Key2', 'A', 'D']

df1 = new dfd.DataFrame(data, { columns: colum1 })
df2 = new dfd.DataFrame(data2, { columns: colum2 })
table(df1)
```

Printing the first DataFrame results in the following output:

|   | Key1 | Key2 | A | B |
|---|------|------|---|---|
| 0 | K0 | K0 | A0 | B0 |
| 1 | K0 | K1 | A1 | B1 |
| 2 | K1 | K0 | A2 | B2 |
| 3 | K2 | K2 | A3 | B3 |

Figure 4.34 – First DataFrame to perform a merge operation on

We will now print the second DataFrame, as follows:

```
table(df2)
```

This results in the following output:

| | Key1 | Key2 | A | D |
|---|---|---|---|---|
| **0** | K0 | K0 | C0 | D0 |
| **1** | K1 | K0 | C1 | D1 |
| **2** | K1 | K0 | C2 | D2 |
| **3** | K2 | K0 | C3 | D3 |

Figure 4.35 – Second DataFrame to perform a merge operation on

Next, we will perform an inner join, as shown in the following code:

```
merge_df = dfd.merge({left: df1, right: df2, on: ["Key1"]})
table(merge_df)
```

Printing the DataFrame results in the following output:

| | Key1 | Key2 | A | B | Key2_1 | A_1 | D |
|---|---|---|---|---|---|---|---|
| **0** | K0 | K0 | A0 | B0 | K0 | C0 | D0 |
| **1** | K0 | K1 | A1 | B1 | K0 | C0 | D0 |
| **2** | K1 | K0 | A2 | B2 | K0 | C1 | D1 |
| **3** | K1 | K0 | A2 | B2 | K0 | C2 | D2 |
| **4** | K2 | K2 | A3 | B3 | K0 | C3 | D3 |

Figure 4.36 – Inner merge of two DataFrames on a single key

From the preceding output, you can see that the inner join in `Key1` results in a DataFrame that contains all the values from DataFrame 1 and DataFrame 2. This is also called a **Cartesian product** of the two DataFrames.

We can take things further and merge multiple keys. We'll look at this in the next section.

## Inner merge on two keys

Performing an inner merge on two keys will also return the intersection of both DataFrames, but it will only return rows with keys that have been found in both the left- and right-hand sides of the DataFrame.

Using the same DataFrames we created earlier, we can perform an inner join on two keys, as shown in the following code:

```
merge_mult_df = dfd.merge({left: df1, right: df2, on: ["Key1",
"Key2"]})
table(merge_mult_df)
```

Printing the DataFrame results in the following output:

| | Key1 | Key2 | A | B | A_1 | D |
|---|---|---|---|---|---|---|
| 0 | K0 | K0 | A0 | B0 | C0 | D0 |
| 1 | K1 | K0 | A2 | B2 | C1 | D1 |
| 2 | K1 | K0 | A2 | B2 | C2 | D2 |

Figure 4.37 – Inner merge of two DataFrames on two keys

From the preceding output, you can see that performing an inner merge on two keys results in a DataFrame with only those keys present in the left- and right-hand sides of the DataFrame.

## Outer merge on a single key

The outer merge, as we saw in the preceding table, performs a union of two DataFrames. That is, it returns values from the keys present in both DataFrames. Performing an outer join on a single key will return the same result as performing an inner join on a single key.

Using the same DataFrames from earlier, we can specify the how parameter to change the merge behavior from its default inner join to outer, as shown in the following code:

```
merge_df = dfd.merge({ left: df1, right: df2, on: ["Key1"],
how: "outer"})
table(merge_df)
```

Printing the DataFrame results in the following output:

| | Key1 | Key2 | A | B | Key2_1 | A_1 | D |
|---|---|---|---|---|---|---|---|
| 0 | K0 | K0 | A0 | B0 | K0 | C0 | D0 |
| 1 | K0 | K1 | A1 | B1 | K0 | C0 | D0 |
| 2 | K1 | K0 | A2 | B2 | K0 | C1 | D1 |
| 3 | K1 | K0 | A2 | B2 | K0 | C2 | D2 |
| 4 | K2 | K2 | A3 | B3 | K0 | C3 | D3 |

Figure 4.38 – Outer merge of two DataFrames on a single key

From the preceding diagram, you can see that the union of the first DataFrame with keys (K0, K0, K1, K2) and the second DataFrame with keys (K0, K1, K1, K2) is (K0, K0, K1, K1, K2). These keys are then used to unionize the DataFrames. After doing this, we received the result shown in the preceding table.

We can also perform an outer join on two keys. This will be different from the inner join on two keys, as we will see in the following sub-section.

## Outer merge on two keys

Still using the same DataFrames from earlier, we can add a second key, as shown in the following code:

```
merge_df = dfd.merge({ left: df1, right: df2, on: ["Key1",
"Key2"], how: "outer"})
table(merge_df)
```

Printing the DataFrame results in the following output:

|   | Key1 | Key2 | A   | B   | A_1 | D   |
|---|------|------|-----|-----|-----|-----|
| 0 | K0   | K0   | A0  | B0  | C0  | D0  |
| 1 | K0   | K1   | A1  | B1  | NaN | NaN |
| 2 | K1   | K0   | A2  | B2  | C1  | D1  |
| 3 | K1   | K0   | A2  | B2  | C2  | D2  |
| 4 | K2   | K2   | A3  | B3  | NaN | NaN |
| 5 | K2   | K0   | NaN | NaN | C3  | D3  |

Figure 4.39 – Outer merge of two DataFrames on two keys

In the preceding output, we can see that the rows that were returned always have keys in Key1 and Key2 present. If the keys present in the first DataFrame are not in the second, the values are always filled with NaN.

> **Note**
> Merging in Danfo.js does not support merging on more than two keys at a time. This may change in the future, so if you need support for such a feature, be sure to check out the *Discussions* section in the official Danfo.js GitHub repository (https://github.com/opensource9ja/danfojs).

In the following sections, we'll briefly look at left and right joins, which are also important when performing merging.

## Right and left merges

Right and left joins are pretty straightforward; they only return rows with keys present in the specified how parameter. For instance, let's say we specify the how parameter as `right`, as shown in the following code:

```
merge_df = dfd.merge({ left: df1, right: df2, on: ["Key1",
"Key2"], how: "right"})
table(merge_df)
```

Printing the DataFrame results in the following output:

|   | Key1 | Key2 | A | B | A_1 | D |
|---|------|------|-----|-----|-----|-----|
| 0 | K0 | K0 | A0 | B0 | C0 | D0 |
| 1 | K1 | K0 | A2 | B2 | C1 | D1 |
| 2 | K1 | K0 | A2 | B2 | C2 | D2 |
| 3 | K2 | K0 | NaN | NaN | C3 | D3 |

Figure 4.40 – Right merge of two DataFrames on two keys

The resulting diagram shows only rows whose keys are present in the right DataFrame. This means that the right DataFrame is given a higher preference when performing the merge.

If we set how to `left`, then a higher preference is given to the left DataFrame, and we only see rows whose keys are present in the left DataFrame. The following code shows an example of performing a `left` join:

```
merge_df = dfd.merge({ left: df1, right: df2, on: ["Key1",
"Key2"], how: "left"})
table(merge_df)
```

Printing the DataFrame results in the following output:

| | Key1 | Key2 | A | B | A_1 | D |
|---|---|---|---|---|---|---|
| 0 | K0 | K0 | A0 | B0 | C0 | D0 |
| 1 | K0 | K1 | A1 | B1 | NaN | NaN |
| 2 | K1 | K0 | A2 | B2 | C1 | D1 |
| 3 | K1 | K0 | A2 | B2 | C2 | D2 |
| 4 | K2 | K2 | A3 | B3 | NaN | NaN |

Figure 4.41 – Left merge of two DataFrames on two keys

Looking at the preceding output, we can confirm that the resulting row gives more preference to the left DataFrame.

The Danfo.js merge function is very powerful and useful and will come in handy when you start working with more than one dataset with overlapping keys. In the next section, we'll introduce another useful function, known as **concatenation**, for transforming datasets.

## Data concatenation

Concatenating data is another important data combination technique, and as its name suggests, this is basically joining, stacking, or arranging data together along an axis.

The function for concatenating data in Danfo.js is exposed as danfo.concat, and the signature of this function is as follows:

```
danfo.concat({df_list, axis})
```

Let's understand what each parameter represents:

- df_list: The df_list parameter is an array of DataFrames or Series you want to join together.

- axis: The axis parameters can accept a row (0) or column (1) and specify the axis where the objects will be aligned.

Next, we'll present some examples that will help you understand how to use the concat function. First up, we will learn how to concatenate three DataFrames along the row axis.

## Concatenating DataFrames along the row axis (0)

First, let's create three DataFrames, as shown in the following code:

```
df1 = new dfd.DataFrame(
    {
        "A": ["A_0", "A_1", "A_2", "A_3"],
        "B": ["B_0", "B_1", "B_2", "B_3"],
        "C": ["C_0", "C_1", "C_2", "C_3"]
    },
    index=[0, 1, 2],
)
df2 = new dfd.DataFrame(
    {
        "A": ["A_4", "A_5", "A_6", "A_7"],
        "B": ["B_4", "B_5", "B_6", "B_7"],
        "C": ["C_4", "C_5", "C_6", "C_7"],
    },
    index=[4, 5, 6],
)
df3 = new dfd.DataFrame(
    {
        "A": ["A_8", "A_9", "A_10", "A_11"],
        "B": ["B_8", "B_9", "B_10", "B_11"],
        "C": ["C_8", "C_9", "C_10", "C_11"]
    },
    index=[8, 9, 10],
)
table(df1)
table(df2)
table(df3)
```

Printing each DataFrame using the print function in Node.js and the browser, or using the table function in Dnotebook, will give us the following output.

The output of the df1 DataFrame is as follows:

| | A | B | C |
|---|---|---|---|
| 0 | A_0 | B_0 | C_0 |
| 1 | A_1 | B_1 | C_1 |
| 2 | A_2 | B_2 | C_2 |
| 3 | A_3 | B_3 | C_3 |

Figure 4.42 – First DataFrame (df1) to be concatenated

The output of the df2 DataFrame is as follows:

| | A | B | C |
|---|---|---|---|
| 0 | A_4 | B_4 | C_4 |
| 1 | A_5 | B_5 | C_5 |
| 2 | A_6 | B_6 | C_6 |
| 3 | A_7 | B_7 | C_7 |

Figure 4.43 – Second DataFrame (df2) to be concatenated

Finally, the output of the df3 DataFrame is as follows:

| | A | B | C |
|---|---|---|---|
| 0 | A_8 | B_8 | C_8 |
| 1 | A_9 | B_9 | C_9 |
| 2 | A_10 | B_10 | C_10 |
| 3 | A_11 | B_11 | C_11 |

Figure 4.44 – Third DataFrame (df3) to be concatenated

Now, let's combine these DataFrames using the concat function, as shown in the following code:

```
df_frames = [df1, df2, df3]
combined_df = dfd.concat({df_list: df_frames, axis: 0})
table(combined_df)
```

This results in the following output:

| | A | B | C |
|---|---|---|---|
| **0_row0** | A_0 | B_0 | C_0 |
| **1_row0** | A_1 | B_1 | C_1 |
| **2_row0** | A_2 | B_2 | C_2 |
| **3_row0** | A_3 | B_3 | C_3 |
| **0_row1** | A_4 | B_4 | C_4 |
| **1_row1** | A_5 | B_5 | C_5 |
| **2_row1** | A_6 | B_6 | C_6 |
| **3_row1** | A_7 | B_7 | C_7 |
| **0_row2** | A_8 | B_8 | C_8 |
| **1_row2** | A_9 | B_9 | C_9 |
| **2_row2** | A_10 | B_10 | C_10 |
| **3_row2** | A_11 | B_11 | C_11 |

Figure 4.45 – Result of concatenating three DataFrames along the row axis (0)

Here, you can see that the concat function simply combines each DataFrame (df1, df2, df3) along the column axis; that is, it stacks them below the first DataFrame in df_list to create one giant combination.

Also, the index looks different. This is because, internally, Danfo.js generated new indexes for the combination. If you need numeric indexes, then you can use the reset_index function, as shown in the following code:

```
combined_df.reset_index(true)
table(combined_df)
```

This results in the following output:

| | A | B | C |
|---|---|---|---|
| 0 | A_0 | B_0 | C_0 |
| 1 | A_1 | B_1 | C_1 |
| 2 | A_2 | B_2 | C_2 |
| 3 | A_3 | B_3 | C_3 |
| 4 | A_4 | B_4 | C_4 |
| 5 | A_5 | B_5 | C_5 |
| 6 | A_6 | B_6 | C_6 |
| 7 | A_7 | B_7 | C_7 |
| 8 | A_8 | B_8 | C_8 |
| 9 | A_9 | B_9 | C_9 |
| 10 | A_10 | B_10 | C_10 |
| 11 | A_11 | B_11 | C_11 |

Figure 4.46 – Resetting the index of the combined DataFrames

Now, what would a concatenation along the column axis using the same DataFrames look like? We'll try this out in the following section.

## Concatenating DataFrames along the column axis (1)

Using the same DataFrames we created earlier, simply change axis to 1 in the concat code, as shown in the following code:

```
df_frames = [df1, df2, df3]
combined_df = dfd.concat({df_list: df_frames, axis: 1})
table(combined_df)
```

Printing the DataFrame results in the following output:

| | A | B | C | D | A_2 | B_2 | C_2 | D_2 | A_3 | B_3 | C_3 | D_3 |
|---|---|---|---|---|---|---|---|---|---|---|---|---|
| 0 | A0 | B0 | C0 | D0 | A4 | B4 | C4 | D4 | A8 | B8 | C8 | D8 |
| 1 | A1 | B1 | C1 | D1 | A5 | B5 | C5 | D5 | A9 | B9 | C9 | D9 |
| 2 | A2 | B2 | C2 | D2 | A6 | B6 | C6 | D6 | A10 | B10 | C10 | D10 |
| 3 | A3 | B3 | C3 | D3 | A7 | B7 | C7 | D7 | A11 | B11 | C11 | D11 |

Figure 4.47 – Result of applying concat to three DataFrames along the column axis (0)

Concatenation also works on Series objects. We'll look at an example in the next section.

## Concatenating Series along a specified axis

Series are also Danfo.js data structures, and as such, the `concat` function can also be used on them. This works the same way as concatenating DataFrames, as we'll demonstrate shortly.

Using the DataFrames from the previous section, we'll create some Series objects, as shown in the following code:

```
series_list = [df1['A'], df2['B'], df3['D']]
```

Notice that we are using a DataFrame sub-setting to grab different columns as Series from the DataFrames. Now, we can combine these Series into an array, which we'll then pass to the `concat` function, as shown in the following code:

```
series_list = [df1['A'], df2['B'], df3['D']]
combined_series = dfd.concat({df_list: series_list, axis: 1})
table(combined_series)
```

Printing the DataFrame results in the following output:

| | A | B | D |
|---|---|---|---|
| 0 | A0 | B4 | D8 |
| 1 | A1 | B5 | D9 |
| 2 | A2 | B6 | D10 |
| 3 | A3 | B7 | D11 |

Figure 4.48 – Result of applying concat to three Series along the row axis (0)

Changing the axis to row (0) will also work and return a DataFrame with lots of missing entries, as shown in the following diagram:

|    | A   | B   | D   |
|----|-----|-----|-----|
| 0  | A0  | NaN | NaN |
| 1  | A1  | NaN | NaN |
| 2  | A2  | NaN | NaN |
| 3  | A3  | NaN | NaN |
| 4  | NaN | B4  | NaN |
| 5  | NaN | B5  | NaN |
| 6  | NaN | B6  | NaN |
| 7  | NaN | B7  | NaN |
| 8  | NaN | NaN | D8  |
| 9  | NaN | NaN | D9  |
| 10 | NaN | NaN | D10 |
| 11 | NaN | NaN | D11 |

Figure 4.49 – Result of applying concat to three Series along the column axis (1)

Now, you may be wondering why there are lots of missing fields in the resulting combination. This is because when objects are combined along a specified axis, sometimes, the position/length of the object will not align with the first DataFrame. Padding is done with NaN to return consistent lengths.

In this sub-section, we covered two important functions (merge and concat) that can help you perform complex combinations on DataFrames or Series. In the next section, we'll talk about something different but important as well, and that is **string/text manipulation**.

# Series data accessors

Danfo.js provides data type-specific methods under various accessors. **Accessors** are namespaces within the Series object that can only be applied/called on specific data types. Two accessors are currently provided for string and date-time Series, and in this section, we'll discuss each and provide some examples for clarity.

## String accessors

String columns in DataFrames or a Series with a `dtype` string can be accessed under the `str` accessor. Calling the `str` accessor on such an object exposes numerous string functions for manipulating the data. We will present some examples in this section.

Let's assume we have a Series that contains the following fields:

```
data = ['lower boy', 'capitals', 'sentence', 'swApCaSe']
sf = new dfd.Series(data)
table(sf)
```

Printing this Series results in the following diagram:

|   | 0 |
|---|---|
| 0 | lower boy |
| 1 | capitals |
| 2 | sentence |
| 3 | swApCaSe |

Figure 4.50 – Result of applying concat to three Series along the column axis (1)

From the preceding output, we can see that the Series (`sf`) contains text and is of the string type. You can confirm this with the `dtype` function, as shown in the following code:

```
console.log(sf.dtype)
```

The output is as follows:

```
string
```

Now that we have our Series, which is of the string data type, we can call the `str` accessor on it and use various JavaScript string methods, such as `capitalize`, `split`, `len`, `join`, `trim`, `substring`, `slice`, `replace`, and so on, as shown in the following examples.

The following code applies the `capitalize` function to a Series:

```
mod_sf = sf.str.capitalize()
table(mod_sf)
```

Printing this Series results in the following output:

| | 0 |
|---|---|
| 0 | Lower boy |
| 1 | Capitals |
| 2 | Sentence |
| 3 | Swapcase |

Figure 4.51 – Result of applying capitalize to a string Series

The following code applies the `substring` function to a Series:

```
mod_sf = sf.str.substring(0,3) //returns a substring by start
and end index
table(mod_sf)
```

Printing this Series results in the following output:

| | 0 |
|---|---|
| 0 | low |
| 1 | cap |
| 2 | sen |
| 3 | swA |

Figure 4.52 – Result of applying the substring function to a string Series

The following code applies the `replace` function to a Series:

```
mod_sf = sf.str.replace("lower", "002") //replaces a string
with specified value
table(mod_sf)
```

Printing this Series results in the following output:

| | 0 |
|---|---|
| 0 | 002 boy |
| 1 | capitals |
| 2 | sentence |
| 3 | swApCaSe |

Figure 4.53 – Result of applying the replace function to a string Series

The following code applies the `join` function to a Series:

```
mod_sf = sf.str.join("7777", "+") // joins specified value to
Series
table(mod_sf)
```

Printing this Series results in the following diagram:

| | 0 |
|---|---|
| 0 | lower boy+7777 |
| 1 | capitals+7777 |
| 2 | sentence+7777 |
| 3 | swApCaSe+7777 |

Figure 4.54 – Result of applying the join function to a string Series

The following code applies the `indexOf` function to a Series:

```
mod_sf = sf.str.indexOf("r")  //Returns the index where the
value is found else -1
table(mod_sf)
```

Printing this Series results in the following output:

| | 0 |
|---|---|
| 0 | 4 |
| 1 | -1 |
| 2 | -1 |
| 3 | -1 |

Figure 4.55 – Result of applying the indexOf function to a string Series

> **Note**
>
> There are numerous exposed string methods available via the `str` accessor. You can see the full list in the Danfo.js documentation (`https://danfo.jsdata.org/api-reference/series#string-handling`).

## Date-time accessors

The second accessor that's exposed by Danfo.js on Series objects is the date-time accessor. This can be accessed under the dt namespace. Processing and extracting different information such as the day, month, and year from a date-time column is a common process when doing data transformations since the date-time raw format is barely useful.

If your data has a date-time column, the dt accessor can be called on it, and this, in turn, exposes various functions that can be used to extract information, such as the year, month, month name, day of the week, hour, second, and minute of the day.

Let's look at some examples of using the dt accessor on a Series with date columns. First, we'll create a Series with some date-time fields:

```
timeColumn = ['12/13/2016 15:00:20', '10/20/2019 18:30:00',
'1/1/2020 12:00:00', '1/30/2020 16:20:00', '11/12/2019
22:00:30']
sf = new dfd.Series(timeColumn, {columns: ["times"]})
table(sf)
```

Printing this Series results in the following output:

|   | times |
|---|-------|
| 0 | 12/13/2016 15:00:20 |
| 1 | 10/20/2019 18:30:00 |
| 2 | 1/1/2020 12:00:00 |
| 3 | 1/30/2020 16:20:00 |
| 4 | 11/12/2019 22:00:30 |

Figure 4.56 – Series with date-time fields

Now that we have a Series with date fields, we can extract some information, such as the hour of the day, year, month, or day of the week. The first thing we need to do is convert the Series into Danfo.js's date-time format using the dt accessor, as shown in the following code:

```
dateTime = sf['times'].dt
```

The `dateTime` variable now exposes different methods for extracting date information. Leveraging this, we can do any of the following:

- Get the hour of the day as a new Series:

```
dateTime = sf.dt
hours = dateTime.hour()
table(hours)
```

This gives us the following output:

|   | 0 |
|---|---|
| 0 | 15 |
| 1 | 18 |
| 2 | 12 |
| 3 | 16 |
| 4 | 22 |

Figure 4.57 – New Series showing the extracted hour of the day

- Get the year as a new Series:

```
dateTime = sf.dt
years = dateTime.year()
table(years)
```

This gives us the following output:

|   | 0 |
|---|---|
| 0 | 2016 |
| 1 | 2019 |
| 2 | 2020 |
| 3 | 2020 |
| 4 | 2019 |

Figure 4.58 – New Series showing the extracted year

- Get the month as a new Series:

```
dateTime = sf.dt
month_name = dateTime.month_name()
table(month_name)
```

This gives us the following output:

| | 0 |
|---|---|
| 0 | Dec |
| 1 | Oct |
| 2 | Jan |
| 3 | Jan |
| 4 | Nov |

Figure 4.59 – New Series showing the extracted name of the month

- Get the day of the week as a new Series:

```
dateTime = sf.dt
weekdays = dateTime.weekdays()
table(weekdays)
```

This gives us the following output:

| | 0 |
|---|---|
| 0 | Tue |
| 1 | Sun |
| 2 | Wed |
| 3 | Thu |
| 4 | Tue |

Figure 4.60 – New Series showing the extracted day of the week

Some other functions that are exposed by the `dt` accessors are as follows:

- `Series.dt.month`: This returns the month as an integer, starting from 1 (January) to 12 (December).

- `Series.dt.day`: This returns the day of the week as an integer, starting from 0 (Monday) to 6 (Sunday).

- `Series.dt.minutes`: This returns the minute of the day as an integer.

- `Series.dt.seconds`: This returns the seconds of the day as an integer.

Congratulations on making it to the end of this section! Data transformation and aggregation are very important aspects of data analysis. Armed with this knowledge, you can properly transform and wrangle your dataset.

In the next section, we'll show you how to calculate descriptive statistics on your datasets using Danfo.js.

# Calculating statistics

Danfo.js comes with some important statistical and mathematical functions. These functions can be used to generate a summary or descriptive statistics of entire DataFrames, as well as a single Series. In datasets, statistics are important because they can give us better insights into data.

In the following sections, we'll use the popular Titanic dataset, which you can download from the following GitHub repository: `https://github.com/PacktPublishing/Building-Data-Driven-Applications-with-Danfo.js/blob/main/Chapter03/data/titanic.csv`.

First, let's load the dataset into Dnotebook using the `read_csv` function, as shown in the following code:

```
var df //using var so df is available to all cells
load_csv("https://raw.githubusercontent.com/plotly/datasets/
master/finance-charts-apple.csv")
.then((data)=>{
  df = data
})
```

The preceding code loads the Titanic dataset from the specified URL and persists it in the `df` variable.

> **Note**
>
> We are using the `load_csv` and `var` declarations here because we are working in Dnotebook. As we have mentioned consistently, you wouldn't use this approach in a browser or Node.js script.

Now, in a new cell, we can print the head of the loaded dataset:

```
table(df.head())
```

Printing the DataFrame results in the following output:

| | Date | AAPL.Open | AAPL.High | AAPL.Low | AAPL.Close | AAPL.Volume | AAPL.Adjusted | dn | mavg |
|---|---|---|---|---|---|---|---|---|---|
| 0 | 2015-02-17 | 127.489998 | 128.880005 | 126.919998 | 127.830002 | 63152400 | 122.905254 | 106.7410523 | 117.9276669 |
| 1 | 2015-02-18 | 127.629997 | 128.779999 | 127.449997 | 128.720001 | 44891700 | 123.760965 | 107.842423 | 118.9403335 |
| 2 | 2015-02-19 | 128.479996 | 129.029999 | 128.330002 | 128.449997 | 37362400 | 123.501363 | 108.8942449 | 119.8891668 |
| 3 | 2015-02-20 | 128.619995 | 129.5 | 128.050003 | 129.5 | 48948400 | 124.510914 | 109.7854494 | 120.7635001 |
| 4 | 2015-02-23 | 130.020004 | 133 | 129.660004 | 133 | 70974100 | 127.876074 | 110.3725162 | 121.7201668 |

Figure 4.61 – The first five rows of the Titanic dataset

We can call the `describe` function on the entire data to quickly get description statistics for all the columns, as shown in the following code:

```
table(df.describe())
```

Printing the DataFrame results in the following output:

| | AAPL.Open | AAPL.High | AAPL.Low | AAPL.Close | AAPL.Volume | AAPL.Adjusted | dn | mavg | up |
|---|---|---|---|---|---|---|---|---|---|
| count | 506 | 506 | 506 | 506 | 506 | 506 | 506 | 506 | 506 |
| mean | 112.935005 | 113.919449 | 111.942017 | 112.958344 | 43178420 | 110.45932 | 107.311386 | 112.739876 | 118.16835 |
| std | 11.28749 | 11.251892 | 11.263687 | 11.244744 | 19852531.300948 | 10.537529 | 11.095804 | 10.595315 | 10.670752 |
| min | 90 | 91.669998 | 89.470001 | 90.339996 | 11475900 | 89.008369 | 85.508858 | 94.047165 | 97.572723 |
| median | 112.889999 | 114.145001 | 111.800003 | 113.025002 | 37474600 | 110.821123 | 107.351628 | 112.79975 | 118.472542 |
| max | 135.669998 | 136.270004 | 134.839996 | 135.509995 | 162206304 | 135.509995 | 127.289261 | 129.845001 | 138.805359 |
| variance | 127.407424 | 126.605063 | 126.870646 | 126.444264 | 394122999055124.75 | 111.039522 | 123.116875 | 112.260701 | 113.864947 |

Figure 4.62 – Output of using the describe function on the Titanic dataset

In the preceding code, we called the `describe` function and then printed the output with `table`. The `describe` function only works on numeric data types, and by default, it will only auto-select columns that are of the `float32` or `int32` type.

> **Note**
>
> By default, all statistical and mathematical computation removes NaN values before the computation is performed.

The `describe` function provides descriptive statistics for the following:

- **count**: This is the total number of rows in a column, excluding `NaN` values.

- **mean**: The average value in a column.

- **Standard Deviation (std)**: The measure of the amount of variation or dispersion of a set of values.

- **Minimum (min)**: The smallest value in a column.

- **median**: The middle value in a column.

- **Maximum (max)**: The largest value in a column.

- **variance**: The measure of the spread of values from the average.

## Calculating statistics by axis

The `describe` function, while quick and easy to apply, is not very helpful when you need to calculate statistics along a specific axis. In this section, we'll introduce some of the most popular methods and show you how to calculate statistics based on a specified axis.

Central tendencies such as mean, mode, and median can be called on a DataFrame with a specified axis. Here, axis 0 represents `rows`, while axis 1 represents `columns`.

In the following code, we are computing the mean, mode, and median of the numeric columns in the Titanic dataset:

```
df_nums = df.select_dtypes(['float32', "int32"]) //select all
numeric dtype columns
console.log(df_nums.columns)
[AAPL.Open,AAPL.High,AAPL.Low,AAPL.Close,AAPL.Volume,AAPL.
Adjusted,dn,mavg,up]
```

In the preceding code, we selected all the numeric columns so that we can apply mathematical operations. Next, we'll call the `mean` function, which, by default, returns the mean across the column axis (1):

```
col_mean = df_nums.mean()
table(col_mean)
```

Printing the DataFrame results in the following output:

|  | 0 |
|---|---|
| AAPL.Open | 112.93500518798828 |
| AAPL.High | 113.91944885253906 |
| AAPL.Low | 111.9420166015625 |
| AAPL.Close | 112.95834350585938 |
| AAPL.Volume | 43178420 |
| AAPL.Adjusted | 110.45932006835938 |
| dn | 107.31138610839844 |
| mavg | 112.73987579345703 |
| up | 118.16835021972656 |

Figure 4.63 – Calling the mean function on numeric columns

The precision shown in the preceding output looks a bit too long. We can round the result to two decimal places to make it more presentable, as shown in the following code:

```
col_mean = df_nums.mean().round(2)
table(col_mean)
```

Printing the DataFrame results in the following output:

|  | 0 |
|---|---|
| AAPL.Open | 112.94 |
| AAPL.High | 113.92 |
| AAPL.Low | 111.94 |
| AAPL.Close | 112.96 |
| AAPL.Volume | 43178420 |
| AAPL.Adjusted | 110.46 |
| dn | 107.31 |
| mavg | 112.74 |
| up | 118.17 |

Figure 4.64 – Calling the mean function on numeric columns and rounding values to two decimal places

The resulting output looks cleaner when rounded to two decimal places. Next, let's calculate mean across the row axis:

```
row_mean = df_nums.mean(axis=0)
table(row_mean)
```

This gives us the following output:

| | 0 |
|---|---|
| 0 | 7017043.5 |
| 1 | 4988077 |
| 2 | 4151488.5 |
| 3 | 5438822.5 |
| 4 | 7886124 |
| 5 | 7692124.5 |
| 6 | 8301412.5 |
| 7 | 10143169 |
| 8 | 6890646.5 |
| 9 | 5344191.5 |

Figure 4.65 – Calling the mean function on the row axis

In the preceding output, you can see that the returned values have numeric labels. This is because the row axis originally had numeric labels.

Using the same idea, we can compute statistics such as mode, median, standard deviation, variance, cumulative sum, cumulative mean, absolute values, and so on. The following are the statistical functions that are currently available in Danfo.js:

- abs: Returns a Series/DataFrame with the absolute numeric value of each element.
- count: Counts the cells for each column or row, excluding NaN values.
- cummax: Returns the cumulative maximum over a DataFrame or Series.
- cummin: Returns the cumulative minimum over a DataFrame or Series.
- cumprod: Returns the cumulative product over a DataFrame or Series.
- cumsum: Returns the cumulative sum over a DataFrame or Series.
- describe: Generates descriptive statistics.
- max: Returns the maximum of the values for the requested axis.
- mean: Returns the mean of the values for the requested axis.
- median: Returns the median of the values for the requested axis.
- min: Returns the minimum of the values for the requested axis.
- mode: Returns the mode(s) of elements along the selected axis.

- `sum`: Returns the sum of the values for the requested axis.

- `std`: Returns the standard deviation over the requested axis.

- `var`: Returns unbiased variance over the requested axis.

- `nunique`: Counts the distinct elements over the requested axis.

> **Note**
> The list of supported functions will likely change, and new ones will
> be added. So, it is better to keep track of the new ones by checking
> out `https://danfo.jsdata.org/api-reference/`
> `dataframe#computations-descriptive-stats`.

Descriptive statistics are very important, and in this section, we successfully covered some important functions that can help you calculate statistics based on a specified axis effectively in Danfo.js.

# Summary

In this chapter, we have successfully covered the data transformation and wrangling functions available in Danfo.js. First, we introduced you to the various wrangling functions for replacing values, filling missing values, and detecting missing values, as well as applying and mapping methods for applying custom functions to your dataset. Knowledge of these functions and techniques ensures you have the required foundation for building data-driven products, as well as getting insights from your data.

Next, we showed you how to use the various merge and concatenation functions in Danfo.js. Finally, we showed you how to calculate descriptive statistics on your dataset.

In the next chapter, we'll take things a step further and show you how to make beautiful and amazing charts/plots using Danfo.js's built-in plotting features, as well as by integrating third-party plotting libraries.

# 5

# Data Visualization with Plotly.js

Plotting and visualization are very important tasks in data analysis, and as such, we are dedicating a full chapter to them. A data analyst will typically perform plotting and data visualization as part of the **exploratory data analysis (EDA)** phase. This can greatly help in identifying useful patterns hidden in data and in building intuition for data modeling.

In this chapter, you will learn how to use **Plotly.js** to create rich and interactive plots that can be embedded into any web application.

Specifically, we'll cover the following topics:

- A brief primer on Plotly.js
- Fundamentals of Plotly.js
- Creating basic charts with Plotly.js
- Creating statistical charts with Plotly.js

# Technical requirements

In order to follow along with this chapter, you should have the following:

- A modern browser such as Chrome, Safari, Opera, or Firefox
- **Node.js** and, optionally, **Danfo Notebook** (**Dnotebook**) installed on your system
- A stable internet connection for downloading datasets

Installation instructions for Dnotebook can be found in *Chapter 2, Dnotebook - An Interactive Computing Environment for JavaScript.*

---

Note

If you do not want to install any software or libraries, you can use the online version of Dnotebook at `https://playnotebook.jsdata.org/`.

---

**Danfo.js** comes with a plotting **application programming interface** (**API**) for easily making plots, and under the hood, it uses Plotly. This is the main reason why we introduce Plotly.js in this chapter, as knowledge gained here will help you easily customize plots created with Danfo.js in the next chapter.

# A brief primer on Plotly.js

Plotly.js (`https://plotly.com/javascript/`), according to the authors, is an open source, high-level, declarative charting library built on top of the popular D3.js (`https://d3js.org/`) and stack.gl (`https://github.com/stackgl`) libraries.

It supports over 40 chart types, including these kinds:

- Basic charts such as scatter plots, line plots, and bar and pie charts
- Statistical graphs such as box plots, histograms, and density plots
- Scientific charts such as heatmaps, log plots, and contour plots
- Financial charts such as waterfall, candlestick, and time-series charts
- Maps such as bubble, choropleth, and Mapbox maps
- **Three-dimensional** (**3D**) charts for scatter plots and surface plots, as well as 3D meshes

To use Plotly.js, you need access to the browser's **Document Object Model** (**DOM**). This means that Plotly.js is a client-side library, and as such can only be used in the browser via a **content delivery network** (**CDN**) script or in client-side libraries such as `React` and `Vue`. In the following section, we will see how to install Plotly.js.

## Using Plotly.js via a script tag

In order to use Plotly.js in a **HyperText Markup Language** (**HTML**) file, you need to add Plotly.js via its `script` tag. In the following code snippet, we add the Plotly.js `script` tag to the header section of a simple HTML file:

```
<!DOCTYPE html>
<html lang="en">
<head>
    <meta charset="UTF-8">
    <meta http-equiv="X-UA-Compatible" content="IE=edge">
    <meta name="viewport" content="width=device-width, initial-
scale=1.0">
    <title>Document</title>
    <script src="https://cdn.plot.ly/plotly-1.2.0.min.js"></
script>
</head>
<body>

</body>
</html>
```

Once you have added the Plotly.js `script` tag as shown in the preceding code snippet, save the HTML file and open it in a browser. The output is going to be an empty page, but under the hood, Plotly.js gets added and is made available in the page. We can test this by making a simple plot, following the steps here:

1.  Create a `div` tag in the HTML body where the graph will be drawn. We'll give this an **identifier** (**ID**) of `myPlot`, as shown here:

    ```
    <body>
      <div id="myPlot">
      </body
    ```

2.  In the body of your HTML page, create sample x and y data, and then plot a
    scatter plot, as shown in the following code snippet:

```
. . .
<body>
    <div id="myPlot"></div>
    <script>
        let data = [{
            x: [1, 3, 5, 6, 8, 9, 5, 8],
            y: [2, 4, 6, 8, 0, 2, 1, 2],
            mode: 'markers',
            type: 'scatter'
        }]
        Plotly.newPlot("myPlot", data)
    </script>
</body>
. . .
```

Opening the HTML file in your browser will give you the following output:

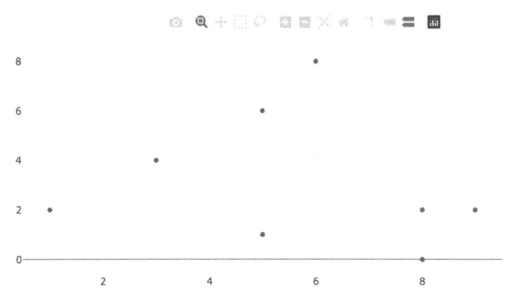

Figure 5.1 – A simple scatter plot made with Plotly

In Dnotebook, which we'll be using a lot in this chapter, you can load and use Plotly by first loading the scripts in a top cell using the `load_package` function, as shown in the following snippet:

```
load_package(["https://cdn.plot.ly/plotly-1.58.4.min.js"])
```

Then, in a new cell, you can add the following code:

```
let data = [{
    x: [1,3,5,6,8,9,5,8],
    y: [2,4,6,8,0,2,1,2],
    mode: 'markers',
    type: 'scatter'
}]


Plotly.newPlot(this_div(), data)
```

Running the preceding code cell will give the following output:

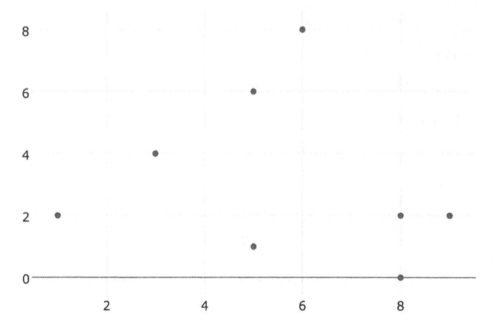

Figure 5.2 – A simple scatter plot made with Plotly on Dnotebook

You can see that the code in the preceding section is the same as the HTML version, with just one slight difference—the `this_div` function passed to `Plotly.newPlot`.

The `this_div` function is just a Dnotebook helper function that creates and returns the ID of the `div` tag just below the code cell block. This means that whenever you're working with plots in Dnotebook, you can get a `div` tag using the `this_div` function.

> **Note**
>
> Going forward, we'll be using `this_div` instead of specifying a `div` tag ID. This is because we'll be working mostly in a Dnotebook environment. To use the code in an HTML or other **user interface (UI)** client, just change `this_div` to the ID of the `div` tag you want to use.

Now you know how to install Plotly, we'll move on to the next section, which is about creating basic charts.

# Fundamentals of Plotly.js

One of the major advantages of using Plotly.js is the fact that it is easy to get started, and there are lots of configurations you can specify to make your plot better. In the section, we are going to cover some of the important configuration options available, and we'll also show you how to specify these options.

Before we go further, let's understand how to get data into Plotly.

## Data format

To make a **two-dimensional (2D)** plot, which is the most common type of plot you'll be creating, you have to pass an object of arrays with x and y keys, as shown in the following code example:

```
const trace1 = {
    x: [20, 30, 40],
    y: [2, 4, 6]
}
```

> **Note**
>
> A data point is normally called a **trace** in Plotly. This is because you can plot more than one data point in a single graph. An example of this is provided here:
>
> ```
> var data = [trace1, trace2]
> Plotly.newPlot("my_div", data);
> ```

The x and y arrays can contain both string and numeric data. If they contain string data, the data points are plotted as they are, that is, point to point. Here's an example of this:

```
var trace1 = {
    x:['2020-10-04', '2021-11-04', '2023-12-04'],
    y: ["cat", "goat", "pig"],
    type: 'scatter'
};
Plotly.newPlot(this_div(),  trace1);
```

Running the preceding code cell will give the following output:

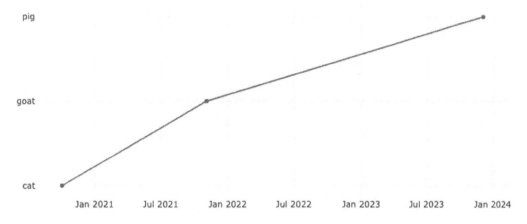

Figure 5.3 – A plot of string values against date with Plotly

If, on the other hand, your data is numeric, then Plotly will automatically sort and then choose a default scale. Look at the following example:

```
var trace1 = {
    x: ['2020-10-04', '2021-11-04', '2023-12-04'],
    y: [90, 20, 10],
    type: 'scatter'
};
var data = [trace1];
Plotly.newPlot(this_div(), data);
```

Running the preceding code cell will give the following output:

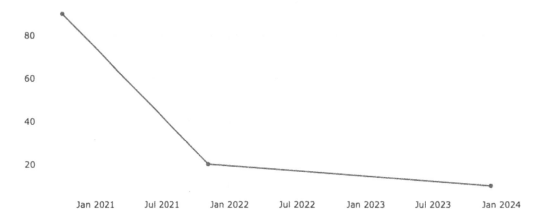

Figure 5.4 – A plot of numerical values against date with Plotly

Before we move to the configuration section, let's see an example of plotting multiple traces in the same chart. First, we set up our data, as shown in the following code snippet:

```
var trace1 = {
    x:['2020-10-04', '2021-11-04', '2023-12-04'],
    y: [90, 20, 10],
    type: 'scatter'
};
var trace2 = {
    x: ['2020-10-04', '2021-11-04', '2023-12-04'],
    y: [25, 35, 65],
    mode: 'markers',
```

```
    marker: {
        size: [20, 20, 20],
    }
};
var data = [trace1, trace2];
Plotly.newPlot(this_div(), data);
```

Running the preceding code cell will give the following output:

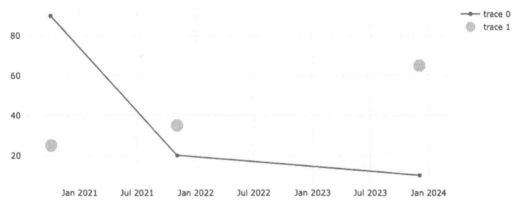

Figure 5.5 – A plot of multiple traces sharing the same x axis

> **Note**
> When plotting multiple traces in a single chart, it is recommended that the traces share a common axis. This makes your plot easier to read.

If you were wondering whether you can add more traces to a data array, then the answer is yes—you can add as many traces as you want, but you must consider interpretability, as adding more traces may not be easy to interpret.

Now you know how to pass data into plots, let's understand some basic configuration options that you can pass to Plotly when making charts.

# Configuration options for plots

Configurations can be used to set properties such as interactivity and modebars of a graph. A configuration is an object and is normally passed as the last argument in the `Plotly.newPlot` call, as shown in the following code snippet:

```
config = { … }
Plotly.newPlot("my_div", data, layout, config)
```

In the following sections, we'll introduce some common configuration options that we will be using in *Chapter 8, Creating a No-Code Data Analysis/Handling System*. If you want to know which configuration options are available to use, then you can read more about it here: `https://plotly.com/javascript/configuration-options/`.

## Configuring the modebar

The **modebar** is a horizontal toolbar that presents numerous options that can be used to interact with a chart. By default, the modebar only becomes visible when you hover over a chart, although this can be changed, as we will see in the following section.

### Making the modebar always visible

To make the modebar always visible, you can set the `displayModeBar` property to `true`, as shown in the following code snippet:

```
var trace1 = {
    x: ['2020-10-04', '2021-11-04', '2023-12-04'],
    y: [90, 20, 10],
    type: 'scatter'
};
var data = [trace1];
var layout = {
    title: 'Configure modebar'
};
var config = {
  displayModeBar: true
};
Plotly.newPlot(this_div(), data, layout, config);
```

Running the preceding code cell will give the following output:

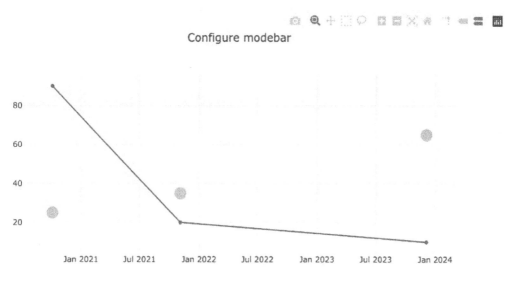

Figure 5.6 – Configuring modebar to always display

If you do not need the modebar, then setting the `displayModeBar` function to `false` will ensure that the modebar is hidden even when you hover over it.

## Removing buttons from the modebar

You can remove buttons from the modebar by passing the names of the buttons you do not want to the `modeBarButtonsToRemove config` property, as we will demonstrate in this section.

Using the same traces as from the *Making the modebar always visible* section, we will remove the zoom-in button from the modebar. You can see the zoomed-in button in the following screenshot before it was removed:

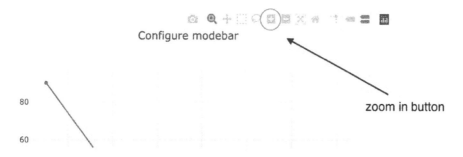

Figure 5.7 – Zoom-in button before removal

In the preceding screenshot, we show the chart before removing the zoom-in button. Next, we'll set the `config` option to remove the button, as shown in the following code snippet:

```
var config = {
    displayModeBar: true,
    modeBarButtonsToRemove: ['zoomIn2d']
};
Plotly.newPlot(this_div(), data, layout, config);
```

Running the preceding code cell will give the following output:

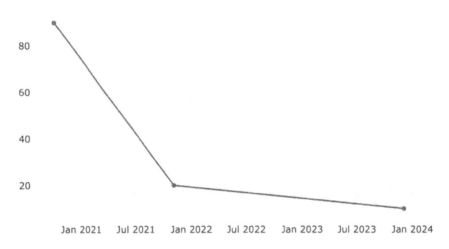

Figure 5.8 – Chart after zoom-in button removal

Using the method demonstrated in the preceding example, you can remove any button from your charts. You can see the names of all modebar buttons that can be removed here: `https://plotly.com/javascript/configuration-options/#remove-modebar-buttons`.

## Adding custom buttons to the modebar

Plotly provides a way to add buttons with custom behavior to the modebar. This becomes useful when we want to extend our plots with custom behaviors— such as, for example, linking to your personal website.

In the following example, we'll add a custom button to display `This is an example of a plot that answers a question on click` to the user when it is clicked.

> **Note**
>
> Adding custom buttons will not work in Dnotebook, so we are going to do this in an HTML file. You can set up an HTML file with a Plotly script, as we demonstrated in the *Using Plotly.js via a script tag* section.

In the body section of your HTML file, add the following code:

```
<div id="mydiv"></div>
<script>
        ...
        var config = {
                displayModeBar: true,
                modeBarButtonsToAdd: [
                        {
                                name: 'about',
                                icon: Plotly.Icons.question,
                                click: function (gd) {
                                        alert('This is an example of a plot
that answers a question on click')
                                }
                        }]
        }
        Plotly.newPlot("mydiv", data, layout, config);
</script>
```

Save and open the preceding HTML file in the browser and click on the button you just created. It should display an alert with the text you specified, similar to the one shown in the following screenshot:

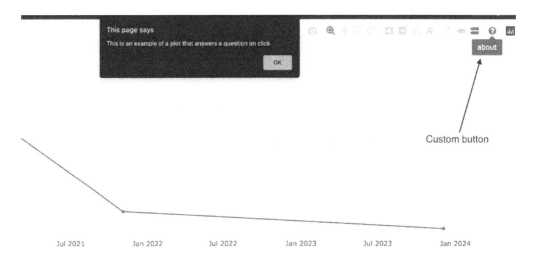

Figure 5.9 – Chart with a custom button

In the code snippet we showed previously, notice the `modeBarButtonsToAdd` configuration option. This option is where we defined the button to add, as well as what happens when we click on it. The main properties you can specify when creating custom buttons are listed here:

- `name`: The name of the button.
- `icon`: The icon/image that is shown in the modebar. This can be a custom icon or any built-in Plotly icon (`https://github.com/plotly/plotly.js/blob/master/src/fonts/ploticon.js`).
- `click`: This defines what happens when you click the button. Here, you can specify any JavaScript function or even change the behavior of the chart.

Next, let's see how to make a static chart.

## Making a static chart

By default, Plotly charts are interactive. If you want to make them static, then you can specify the following option in the `config` object:

```
var config = {
   staticPlot: true
}
```

Static charts can be useful when we only want to display a chart with a distracting interaction.

Next, we'll show you how to create charts that are responsive.

## Making a responsive chart

To make a chart responsive, such that it can automatically resize alongside the window where it is displayed, you can set the `responsive` property to `true`, as shown in the following code snippet:

```
var config = {
    responsive: true
}
```

Responsive charts are useful when you're creating web pages that will be shown across different screen sizes.

In the next section, we'll show you how to download and set download options in your charts.

## Customizing download plot options

By default, when the modebar is shown, you can save a Plotly chart as a **Portable Network Graphics** (**PNG**) file. This can be customized, and you can set the download image type, as well as other properties such as filename, height, width, and so on.

To achieve this, you can set the `toImageButtonOptions` property in the `config` object, as we demonstrate in the following code snippet:

```
var config = {
   toImageButtonOptions: {
      format: 'jpeg', // one of png, svg, jpeg, webp
      filename: 'my_image', // the name of the file
      height: 600,
```

```
    width: 700,
  }
}
```

And finally, in the next section, we'll demonstrate how to change the locale of your charts.

## Changing the default locale

A locale is important when making charts for people speaking other languages. This can greatly improve the interpretability of your charts.

Following these next steps, we'll change the default locale from English to French:

1. Get the specific locale and add it to your HTML file (or load it using `load_scripts` in Dnotebook), as shown in the following code snippet:

```
...
<head>
    <script src="https://cdn.plot.ly/plotly-1.58.4.min.js"></script>
    <script src="https://cdn.plot.ly/plotly-locale-fr-latest.js"></script>  <!-- load locale -->
</head>
...
```

   In Dnotebook, this can be done using `load_package`, as follows:

```
load_package(["https://cdn.plot.ly/plotly-1.58.4.min.js",
"https://cdn.plot.ly/plotly-locale-fr-latest.js"])
```

2. In your `config` object, specify the locale, as shown in the following code snippet:

```
var config = {
    locale: "fr"
}
```

Let's see a complete example with the corresponding output. Add the following code to the body of your HTML file:

```
<div id="mydiv"></div>
  <script>
      var trace1 = {
          x: ['2020-10-04', '2021-11-04', '2023-12-04'],
```

```
        y: [90, 20, 10],
        type: 'scatter'
    };
    var trace2 = {
        x: ['2020-10-04', '2021-11-04', '2023-12-04'],
        y: [25, 35, 65],
        mode: 'markers',
        marker: {
            size: [20, 20, 20],
        }
    };
    var data = [trace1, trace2];
    var layout = {
        title: 'Change Locale',
        showlegend: false
    };
    var config = {
        locale: "fr"
    };
    Plotly.newPlot("mydiv", data, layout, config);
</script>
```

Loading the HTML page in the browser displays the following chart, with `locale` set to French:

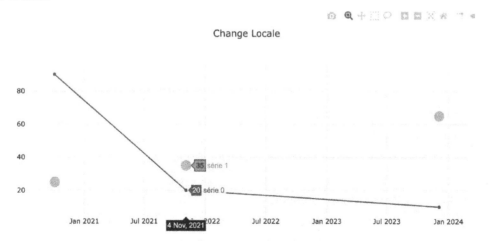

Figure 5.10 – Chart with locale set to French

Now you know how to configure your plots, we'll move on to another important aspect of chart configuration: **layout**.

# Plotly layout

layout (`https://plotly.com/javascript/reference/layout/`) is the third argument passed to the `Plotly.newPlot` function. It is used to configure the area/layout where a chart is drawn, as well as properties such as the title, text, legend, and so on.

There are six layout properties that you can set—title, legend, margins, size, fonts, and colors. We'll demonstrate how to use them, with examples.

## Configuring chart title

The `title` property configures the chart title, which is the text shown at the top of a chart. To add a title, simply pass the text to the `title` property in the `layout` object, as demonstrated in the following code snippet:

```
var layout = {
    title: 'This is an example title,
};
```

To be more explicit, especially if you need to configure how the title text is positioned, then you can set the `text` property of `title`, as we show in the following code snippet:

```
var layout = {
    title: {text: 'This is an example title'}
};
```

Using the preceding format, we can easily configure other properties such as title position using other properties, as outlined here:

- x: A number between 0 and 1 inclusive and used to set the x position of the title text with respect to the container where it's displayed.

- y: Also a number between 0 and 1 inclusive and used to set the y position of the title text with respect to the container where it's displayed.

- xanchor: This can be one of `auto`, `left`, `center`, or `right` alignment. It sets the title's horizontal alignment with respect to its x position.

- yanchor: This can be one of `auto`, `top`, `middle`, or `bottom` alignment. It sets the title's vertical alignment with respect to its y position.

Let's see an example of configuring `title` to show it in the top-right corner of a chart, as follows:

```
var trace1 = {
     x:['2020-10-04', '2021-11-04', '2023-12-04'],
     y: [90, 20, 10],
     type: 'scatter'
};
var data = [trace1];
var layout = {
  title: { text: 'This is an example title',
          x: 0.8,
          y: 0.9,
          xanchor: "right"}
};
Plotly.newPlot(this_div(), data, layout, config);
```

This gives the following output:

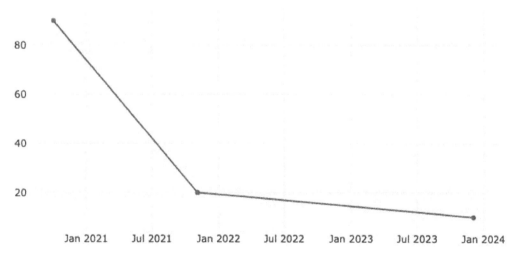

Figure 5.11 – Chart with title configured to top right

You can also set the **padding** of the title. The padding can accept the following parameters:

- `t`: Sets the top padding of the title
- `b`: Sets the bottom padding of the title
- `r`: Sets the right padding and will only work when the `xanchor` property is set to `right`
- `l`: Sets the left padding and will only work when the `xanchor` property is set to `left`

For example, to set the `right` padding of the title, you can set the `xanchor` property to `right` and then configure the `r` property of `pad`, as shown in the following code snippet:

```
var layout = {
  title: {text: 'This is an example title',
          xanchor: "right",
          pad: {r: 100}}
};
```

Note that the padding parameters are in **pixels (px)**, so `100` in the preceding code snippet means 100px.

In the next section, we will take a look at configuring the legend of a chart.

## Configuring Plotly legends

A legend describes the data displayed in a chart. Legends are very important when you're displaying more than one form of data in a single chart.

By default, Plotly shows a legend when you have more than one trace in a chart. You can also explicitly show a legend by setting the `showLegend` property of `layout` to `true`, as illustrated in the following code snippet:

```
var layout = {
  showLegend: true
};
```

Once the legend is activated, you can customize how it is displayed by setting the following properties:

- `bgcolor`: Sets the background color of the legend. By default, it is set to `#fff` (white).

- `bordercolor`: Sets the color of the border enclosing the legend.

- `borderwidth`: Sets the width (in px) of the border enclosing the legend.

- `font`: An object with the following properties:

  a) `family`: Any supported HTML font family.

  b) `size`: The size of the legend text.

  c) `color`: The color of the legend text.

In the following code snippet, we show an example of using these properties to configure your legend:

```
var trace1 = {
     x: ['2020-10-04', '2021-11-04', '2023-12-04'],
     y: [90, 20, 10],
     type: 'scatter'
};
var trace2 = {
     x: ['2020-10-04', '2021-11-04', '2023-12-04'],
     y: [25, 35, 65],
     mode: 'markers',
     marker: {
          size: [20, 20, 20],
     }
};
var data = [trace1, trace2];
var layout = {
  title: {text: 'This is an example title'},
  showLegend: true,
  legend: {bgcolor: "#fcba03",
           bordercolor: "#444",
           borderwidth: 1,
           font: {family: "Arial", size: 30, color: "#fff"}}
};
Plotly.newPlot(this_div(), data, layout, config);
```

The preceding code gives the following output:

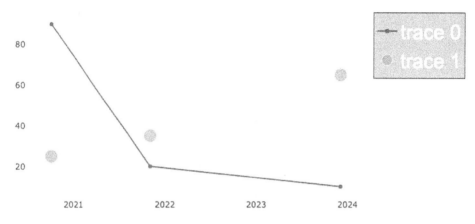

Figure 5.12 – Chart showing legend with custom configuration

Next, we show how to configure the margin of the overall layout.

## Configuring layout margins

The `margin` property configures the position of the chart on the screen. It supports all margin properties (`l`, `r`, `t`, and `b`). In the following code snippet, we use all four properties to demonstrate setting layout margins:

```
...
var layout = {
    title: {text: 'This is an example title'},
    margin: {l: 200, r: 200, t: 230, b: 100}
};
Plotly.newPlot(this_div(), data, layout, config);
```

This gives the following output:

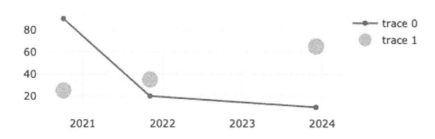

Figure 5.13 – Chart with margin configured

Notice the space around the chart in the preceding screenshot? That is the set margin. It is important to note that the margin is in px as well.

Next, we'll look at setting the layout size.

## Configuring layout size

Sometimes, we may want a bigger or a smaller layout, and this can be configured using the width, height, or—conveniently— the autosize property, as we demonstrate in the following code snippet:

```
...
var layout = {
    title: {text: 'This is an example title'},
    width: 1000,
    height: 500
};
Plotly.newPlot(this_div(), data, layout, config);
```

This gives the following output:

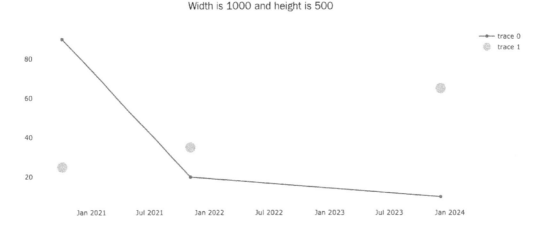

Figure 5.14 – Chart with size configured

The `autosize` property can be set to `true` when we want Plotly to automatically set the size of the layout.

> **Note**
>
> To see other properties that can be configured in `layout`, you can visit Plotly's official API reference here: `https://plotly.com/javascript/reference/layout`.

In the next section, we'll show you how to make different types of charts based on the information you want to convey.

# Creating basic charts with Plotly.js

Plotly.js supports many basic charts that can be quickly used to convey information. Some examples of basic charts available in Plotly are scatter plots, line, bar, pie and bubble charts, dot plots, treemaps, tables, and so on. You can find a complete list of supported basic charts here: `https://plotly.com/javascript/basic-charts/`.

In this section, we will cover some basic charts such as scatter plots, bar charts, and bubble charts.

First, we'll start with scatter plots.

## Creating a scatter plot with Plotly.js

A **scatter plot** is typically used to plot two variables against each other. The plot is displayed as a collection of points, hence the name *scatter plot*. The following screenshot shows an example of a scatter plot:

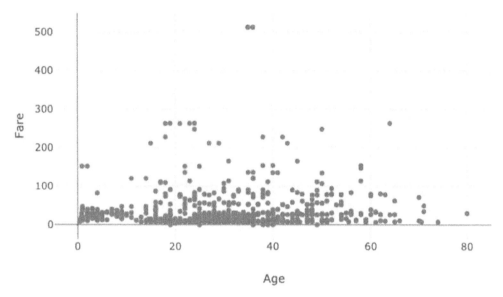

Figure 5.15 – Example of a scatter plot showing Fare against Age margin

To make a scatter plot using Plotly, you simply specify the plot type, as we show in the following example:

```
var trace1 = {
    x: [2, 5, 7, 12, 15, 20],
    y: [90, 80, 10, 20, 30, 40],
    type: 'scatter'
};
var data = [trace1];
Plotly.newPlot(this_div(), data);
```

This gives the following output:

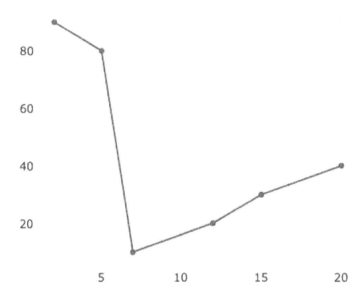

Figure 5.16 – Example scatter plot of sales versus margin

By default, the points are joined together using a line. You can change this behavior by setting a mode type. The mode type can be any of the following:

- `markers`
- `lines`
- `text`
- `none`

You can also use more than one mode by joining them together with a plus sign—for example, `markers+text+lines` or `markers+lines`.

In the following example, we use both `markers` and `text` as our mode type:

```
var trace1 = {
    x: [2, 5, 7, 12, 15, 20],
    y: [90, 80, 10, 20, 30, 40],
    type: 'scatter',
    mode: 'markers+text'
```

```
};
var data = [trace1];
Plotly.newPlot(this_div(), data);
```

This gives the following output:

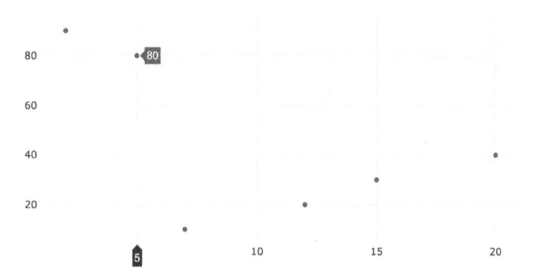

Figure 5.17 – Scatter plot with mode type set to markers+text

As mentioned earlier, you can plot more than one scatter plot in a single chart, and each trace can be configured as required. In the following example, we plot three scatter plots with varying modes configured:

```
var trace1 = {
    x: [1, 2, 3, 4, 5],
    y: [1, 6, 3, 6, 1],
    mode: 'markers',
    type: 'scatter',
    name: 'Trace 1',
};
var trace2 = {
    x: [1.5, 2.5, 3.5, 4.5, 5.5],
    y: [4, 1, 7, 1, 4],
    mode: 'lines',
    type: 'scatter',
    name: 'Trace 2',
```

```
};
var trace3 = {
    x: [1, 2, 3, 4, 5],
    y: [4, 1, 7, 1, 4],
    mode: 'markers+text',
    type: 'scatter',
    name: 'Trace 3',
};
var data = [ trace1, trace2, trace3];
var layout = {
    title:'Data Labels Hover',
    width: 1000
};
Plotly.newPlot(this_div(), data, layout);
```

Running the preceding code cell gives the following output:

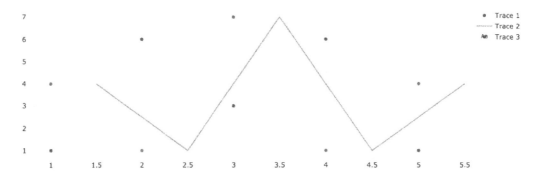

Figure 5.18 – Scatter plot with three traces

Now you have learned the concept of basic charts in this section, you can easily create scatter plots from custom data points and customize the size using the required properties.

Next, we'll briefly look at bar charts.

## Creating bar charts

A bar chart is another popular type of chart available in Plotly.js. It is used to show the relationship between data points using a rectangular bar, with heights or lengths proportional to the values they represent. Bar charts are mostly used for plotting **categorical data**.

> **Note**
>
> Categorical data or a categorical variable is a variable with a fixed or limited number of possible values. The letters of the English alphabet are an example of categorical data.

In order to make a bar chart in Plotly.js, you pass a categorical data point with corresponding bar heights/length and then set the type to `bar`, as shown in the following example:

```
var data = [
  {
    x: ['Apple', 'Mango', 'Pear', 'Banana'],
    y: [20, 20, 15, 40],
    type: 'bar'
  }
];
Plotly.newPlot(this_div(), data);
```

Running the preceding code cell gives the following output:

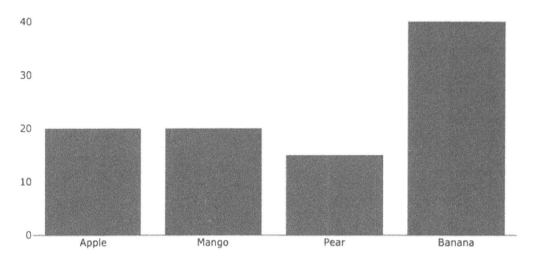

Figure 5.19 – Simple bar chart with four variables

You can plot more than one bar chart in a single layout by creating multiple traces and passing them as an array. For example, in the following code snippet, we create two traces and a plot:

```
var trace1 = {
    x: ['Apple', 'Mango', 'Pear', 'Banana'],
    y: [20, 20, 15, 40],
    type: 'bar'
}
var trace2 = {
    x: ['Goat', 'Lion', 'Spider', 'Tiger'],
    y: [25, 10, 14, 36],
    type: 'bar'
}
var data = [trace1, trace2]
Plotly.newPlot(this_div(), data);
```

Running the preceding code cell gives the following output:

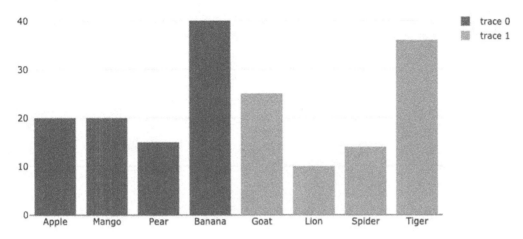

Figure 5.20 – A bar chart with two traces

When plotting multiple traces within the same category, you can specify a `barmode` property. The `barmode` property can be one of `stack` or `group` mode. For example, in the following code snippet, we make a bar chart of two traces with the same categories in `stack` mode:

```
var trace1 ={
    x: ['Apple', 'Mango', 'Pear', 'Banana'],
    y: [20, 20, 15, 40],
    type: 'bar',
    name: "Bar1"
}
var trace2 = {
    x: ['Apple', 'Mango', 'Pear', 'Banana'],
    y: [25, 10, 14, 36],
    type: 'bar',
    name: "Bar2"
}
var data = [trace1, trace2]
var layout = {
    barmode: 'stack'
}
Plotly.newPlot(this_div(), data, layout);
```

Running the preceding code cell gives the following output:

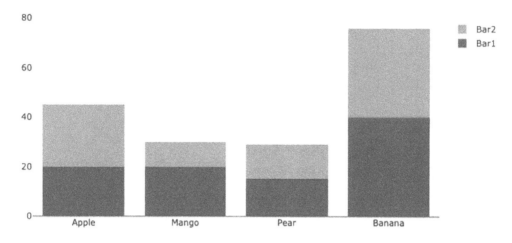

Figure 5.21 – A bar chart with two traces in stack mode

In the following code snippet, we change the barmode property to group (default mode):

```
...
var layout = {
    barmode: 'group'
}
...
```

This results in the following output:

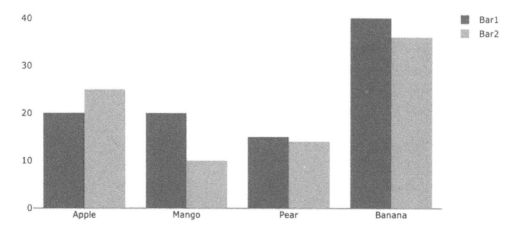

Figure 5.22 – A bar chart with two traces in group mode

There are many other options you can specify when making bar charts, but we will not cover all of them in this section. You can see all configuration options, as well as clear examples of creating good bar charts, in the official documentation here: `https://plotly.com/javascript/bar-charts/`.

In the following section, we'll briefly cover bubble charts.

## Creating bubble charts

A bubble chart is another very popular type of chart that can be used to cover information. It is basically an extension of a scatter plot, with point sizes specified. Let's look at the following code example:

```
var trace1 = {
    x: [1, 2, 3, 4],
    y: [10, 11, 12, 13],
    mode: 'markers',
    marker: {
        size: [40, 60, 80, 100]
    }
};
var data = [trace1]
Plotly.newPlot(this_div(), data, layout);
```

Running the preceding code cell gives the following output:

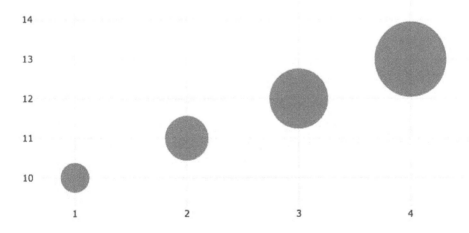

Figure 5.23 – A simple bubble chart

In the previous code snippet for a bubble chart, you can see that the main changes are the mode and the markers with sizes specified. The sizes are mapped one-to-one with the points, and if you want to apply sizes to each bubble, you must specify the size.

You can also change the colors of individual bubbles by passing an array of colors, as demonstrated in the following code snippet:

```
...
  marker: {
    size: [40, 60, 80, 100],
    color: ['rgb(93, 164, 214)', 'rgb(255, 144, 14)', 'rgb(44,
 160, 101)', 'rgb(255, 65, 54)'],
  }
...
```

Running the preceding code cell gives the following output:

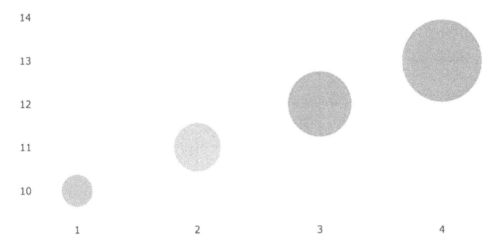

Figure 5.24 – A simple bubble chart with different colors

Bubble charts are very useful, and if you need to know more or see some more advanced examples, you can check out the example page in Plotly's documentation here: https://plotly.com/javascript/bubble-charts/.

There are many other types of basic plots you can make, but sadly we can't cover all of them. The *Basic Charts* page (`https://plotly.com/javascript/basic-charts/`) in Plotly's documentation is a good place to learn how to make these awesome plots, and we encourage you to take a look.

In the next section, we'll introduce you to some statistical charts.

# Creating statistical charts with Plotly.js

Statistical charts are different types of charts used mostly by statisticians or data scientists to convey information. Some examples of statistical plots are histograms, box plots, violin plots, density plots, and so on. In the following sub-section, we'll briefly cover three types of statistical plots—histograms, box plots, and violin plots.

## Creating histogram plots with Plotly.js

A histogram is used to represent the distribution or spread of numerical/continuous data. A histogram is similar to a bar chart, and sometimes people may confuse the two. A simple way to differentiate between them is the type of data they can show. A histogram works with continuous variables instead of categorical variables, and only needs a single value as data.

In the following code snippet, we show an example of a histogram with generated random numbers:

```
var x = [];
for (let i = 0; i < 1000; i ++) { //generate random numbers
x[i] = Math.random();
}
var trace = {
    x: x,
    type: 'histogram',
  };
var data = [trace];
Plotly.newPlot(this_div(), data);
```

In the preceding code snippet, you will observe that the `trace` property has only the x data specified. This is in line with what we mentioned earlier—histograms only need a single value. We also specify the plot type to be `histogram`, and running the code cell gives the following output:

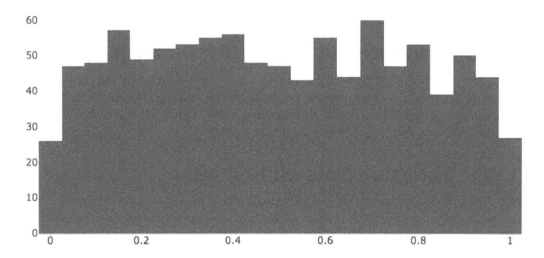

Figure 5.25 – A histogram with random values of x

Specifying a y value instead of x will result in a horizontal histogram, as shown in the following example:

```
...
var trace = {
    y: y,
    type: 'histogram',
};
...
```

Running the preceding code cell gives the following output:

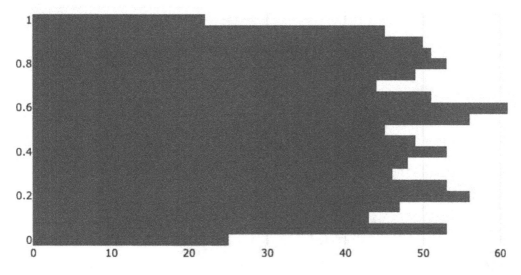

Figure 5.26 – A histogram with random values of y

You can also create stacked, overlaid, or grouped histograms by creating multiple traces and setting the `barmode` property to `stack`, as demonstrated in the following example:

```
var x1 = [];
var x2 = [];
for (var i = 0; i < 1000; i ++) {
x1[i] = Math.random();
x2[i] = Math.random();
}
var trace1 = {
    x: x1,
    type: "histogram",
};
var trace2 = {
    x: x2,
    type: "histogram",
};
```

```
var data = [trace1, trace2];
var layout = {barmode: "stack"};
Plotly.newPlot(this_div(), data, layout);
```

Running the preceding code cell gives the following output:

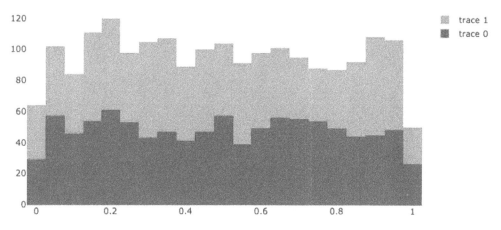

Figure 5.27 – A histogram in stack mode

By changing the `barmode` overlay, we get the following output:

```
...
var layout = {barmode: "overlay"};
...
```

Running the preceding code cell gives the following output:

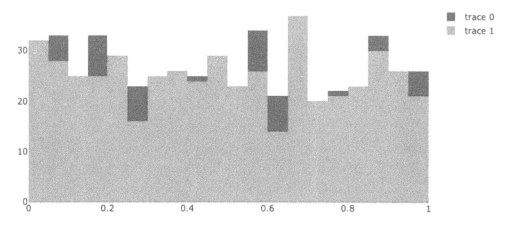

Figure 5.28 – A histogram in overlay mode

To see more examples of plotting histograms as well as various configuration options, you can check out the histogram documentation page here: `https://plotly.com/javascript/histograms/`.

In the next section, we'll introduce box plots.

## Creating box plots with Plotly.js

A **box plot** is a very common type of plot in descriptive statistics. It graphically presents groups of numerical data using their quartiles. Box plots also have lines extending above or below them, called **whiskers**. The whiskers represent the variability outside the upper and lower **quartiles**.

> Tip
> A quartile divides a specified number of data points into four parts or quarters. The first quartile is the lowest 25% of data points, the second quartile is between 25% and 50% (up to the median), the third quartile is 50% to 75% (above the median), and finally, the fourth quartile depicts the highest 25% of numbers.

The following diagram can help you better understand box plots:

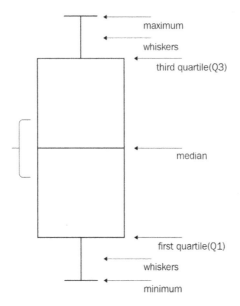

Figure 5.29 – Diagram depicting a box plot (source: Redrawn from https://aiaspirant.com/box-plot/)

In Plotly.js, we make a box plot by passing our data and setting the `trace` type to `box`. We demonstrate this in the following example:

```
var y0 = [];
var y1 = [];
for (var i = 0; i < 50; i ++) {//generate some random numbers
y0[i] = Math.random();
y1[i] = Math.random() + 1;
}
var trace1 = {
    y: y0,
    type: 'box'
};
var trace2 = {
    y: y1,
    type: 'box'
};
var data = [trace1, trace2];
Plotly.newPlot(this_div(), data);
```

Running the preceding code cell gives the following output:

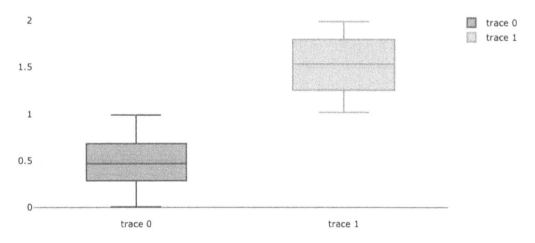

Figure 5.30 – A simple box plot with two traces

We can configure the layout for box plots to be horizontal as opposed to the default vertical format. In the following section, we demonstrate how to do this.

## Making horizontal box plots

You can make horizontal plots by specifying x values instead of y values in your traces. We demonstrate this in the following code snippet:

```
var x0 = [];
var x1 = [];
for (var i = 0; i < 50; i ++) {
x0[i] = Math.random();
x1[i] = Math.random() + 1;
}
var trace1 = {
    x: x0,
    type: 'box'
};
var trace2 = {
    x: x1,
    type: 'box'
};
var data = [trace1, trace2];
Plotly.newPlot(this_div(), data);
```

Running the preceding code cell gives the following output:

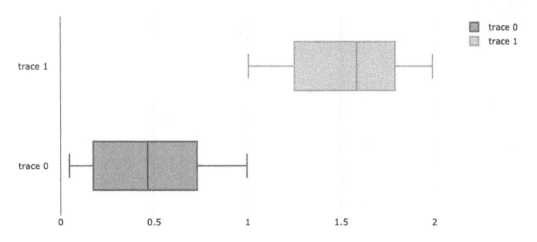

Figure 5.31 – A simple box plot with two traces

You can also make grouped box plots, as we show in the following section.

## Making grouped box plots

Multiple traces sharing the same *x* axis can be grouped together into a single box plot, as shown in the following code snippet:

```
var x = ['Season 1', 'Season 1', 'Season 1', 'Season 1',
'Season 1', 'Season 1',
          'Season 2', 'Season 2', 'Season 2', 'Season 2',
'Season 2', 'Season 2']
var trace1 = {
   y: [2, 2, 6, 1, 5, 4, 2, 7, 9, 1, 5, 3],
   x: x,
   name: 'Blues FC',
   marker: {color: '#3D9970'},
   type: 'box'
};
var trace2 = {
   y: [6, 7, 3, 6, 0, 5, 7, 9, 5, 8, 7, 2],
   x: x,
   name: 'Reds FC',
   marker: {color: '#FF4136'},
   type: 'box'
};
var trace3 = {
   y: [1, 3, 1, 9, 6, 6, 9, 1, 3, 6, 8, 5],
   x: x,
   name: 'Greens FC',
   marker: {color: '#FF851B'},
   type: 'box'
};
var data = [trace1, trace2, trace3];
var layout = {
   yaxis: {
     title: 'Points in two seasons',
   },
   boxmode: 'group'
};
Plotly.newPlot(this_div(), data, layout);
```

Running the preceding code cell gives the following output:

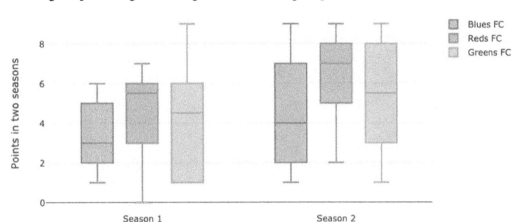

Figure 5.32 – A box plot with three traces grouped together

There are numerous other options you can set when making box plots, but we'll leave you to read more about them in the box-plot documentation here: `https://plotly.com/javascript/box-plots/`.

In the next section, we'll briefly introduce violin plots.

## Creating violin plots with Plotly.js

A **violin plot** is an extension of a box plot. It also describes data points using quartiles, just like a box plot, with just one major difference—the fact that it also shows the distribution of the data.

The following diagram shows the common characteristics between a violin plot and a box plot:

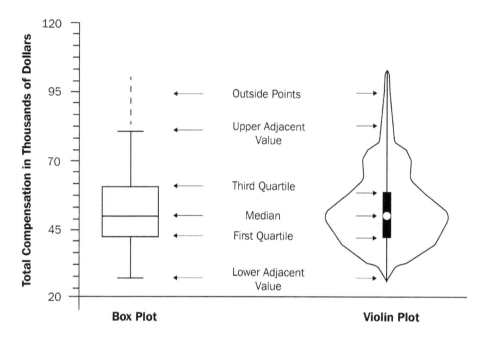

Figure 5.33 – Common properties between a violin plot and a box plot (Redrawn from https://towardsdatascience.com/violin-plots-explained-fb1d115e023d)

The curved area of the violin plot shows the underlying distribution of the data and conveys more information than the box plot.

In Plotly, you can easily make a violin plot by just changing the type to `violin`. For example, in the following code snippet, we are reusing the code from the box-plot section, with just two main changes:

```
var x = ['Season 1', 'Season 1', 'Season 1', 'Season 1',
'Season 1', 'Season 1',
        'Season 2', 'Season 2', 'Season 2', 'Season 2',
'Season 2', 'Season 2']
var trace1 = {
  y: [2, 2, 6, 1, 5, 4, 2, 7, 9, 1, 5, 3],
  x: x,
  name: 'Blues FC',
  marker: {color: '#3D9970'},
```

```
      type: 'violin'
};
var trace2 = {
    y: [6, 7, 3, 6, 0, 5, 7, 9, 5, 8, 7, 2],
    x: x,
    name: 'Reds FC',
    marker: {color: '#FF4136'},
    type: 'violin'
};
var trace3 = {
    y: [1, 3, 1, 9, 6, 6, 9, 1, 3, 6, 8, 5],
    x: x,
    name: 'Greens FC',
    marker: {color: '#FF851B'},
    type: 'violin',
};
var data = [trace1, trace2, trace3];
var layout = {
    yaxis: {
        title: 'Points in two seasons',
    },
    violinmode: 'group'
};
Plotly.newPlot(this_div(), data, layout);
```

Running the preceding code cell gives the following output:

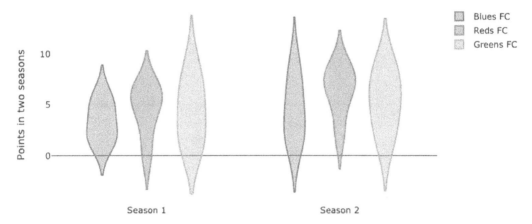

Figure 5.34 – A violin plot with three traces grouped together

Just as with other chart types, you can also configure how violin plots are displayed. For example, we can show the underlying box plot in a violin plot, as shown in the following code snippet:

```
var x = ['Season 1', 'Season 1', 'Season 1', 'Season 1',
'Season 1', 'Season 1',
         'Season 2', 'Season 2', 'Season 2', 'Season 2',
'Season 2', 'Season 2']
var trace = {
  y: [1, 3, 1, 9, 6, 6, 9, 1, 3, 6, 8, 5],
  x: x,
  name: 'Greens FC',
  marker: {color: '#FF851B'},
  type: 'violin',
  box: {
    visible: true
  },
};
var data = [trace];
var layout = {
  yaxis: {
```

```
        title: 'Point in two seasons',
    },
};
Plotly.newPlot(this_div(), data, layout);
```

Running the preceding code cell gives the following output:

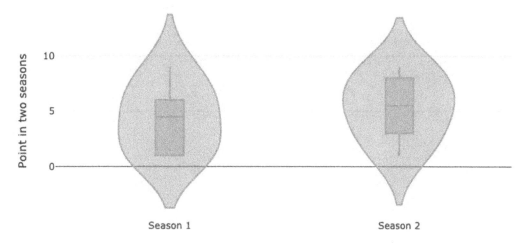

Figure 5.35 – A violin plot with underlying box plot shown

To see other configuration options as well as some advanced settings, you can check out the documentation of a violin plot here: `https://plotly.com/javascript/violin/`.

With that, we have to come to the end of this introductory section on Plotly.js. In the next chapter, we will show you how to use Danfo.js to quickly and easily make plots for any type of data supported by this particular library.

# Summary

In this chapter, we covered plotting and visualization with Plotly.js. First, we gave a brief introduction to Plotly.js, including installation setup. Then, we moved on to chart configuration and layout customization. Finally, we showed you how to create some basic and statistical charts.

The knowledge you have gained in this chapter will help you easily create interactive charts that you can embed in your websites or web applications.

In the next chapter, we'll introduce data visualization with Danfo.js, and you'll see how, with knowledge of Plotly.js, you can easily create amazing charts directly from your DataFrame or Series.

# 6
# Data Visualization with Danfo.js

In the preceding chapter, you learned how to use Plotly.js to create rich and interactive plots that can be embedded in any web application. We also mentioned how Danfo.js uses Plotly internally to make plots directly on a DataFrame or Series. In this chapter, we'll show you how to easily create these charts using the Danfo.js plotting API. Specifically, in this chapter, we'll cover the following:

- Setting up Danfo.js for plotting
- Creating line charts with Danfo.js
- Creating scatter plots with Danfo.js
- Creating box and violin plots with Danfo.js
- Creating histograms with Danfo.js
- Creating bar charts with Danfo.js

# Technical requirements

In order to follow along with this chapter, you should have the following:

- A modern browser such as Chrome, Safari, Opera, or Firefox
- Node.js, Danfo.js, and optionally Dnotebook installed on your system
- A stable internet connection for downloading the datasets

Installation instructions for Danfo.js can be found in *Chapter 3, Getting Started with Danfo.js*, while installation steps for Dnotebook can be found in *Chapter 2, Dnotebook – An Interactive Computing Environment for JavaScript.*

# Setting up Danfo.js for plotting

By default, Danfo.js provides some basic chart types. These charts can be called on any DataFrame or Series object and, if the correct arguments are passed, it will display the corresponding chart.

At the time of writing, Danfo.js comes with the following charts:

- Line charts
- Box and violin plots
- Tables
- Pie charts
- Scatter plots
- Bar charts
- Histograms

These charts are exposed via the `plot` function. That is, if you have a DataFrame or Series object, calling the `plot` function on them exposes these charts.

The `plot` method requires a `div` ID where the plot is to be shown. For example, assuming `df` is a DataFrame, we can call the `plot` function as shown in the following code snippet:

```
const df = new DataFrame({...})
df.plot("my_div_id").<chart type>
```

The chart type can be `line`, `bar`, `scatter`, `hist`, `pie`, `box`, `violin`, or `table`.

Each plot type will accept plot-specific arguments, but they all share a common argument called `config`. The `config` object is used to customize the plot as well as the layout. Think of the `config` argument as a combination of the layout and configuration properties we used in *Chapter 5, Data Visualization with Plotly.js*.

The `config` argument is an object with the following format:

```
config = {
   layout: {…}, // plotly layout parameters like title, font,
e.t.c.
   ... // other Plotly configuration parameters like showLegend,
displayModeBar, e.t.c.
}
```

In the following section, we will show some examples of using the different chart types, as well as showing you how to configure them.

> **Note**
> We'll be downloading and using two real-world datasets in the following sections. This means you'll need an internet connection to download the datasets.

## Adding Danfo.js to your code

To use Danfo.js for plotting, you need to add it to your project. If you're working in Dnotebook, which we'll be using for our examples, then use the `load_package` function to load Danfo.js and Plotly.js as shown in the following code snippet:

```
load_package(["https://cdn.plot.ly/plotly-1.58.4.min.
js","https://cdn.jsdelivr.net/npm/danfojs@0.2.3/lib/bundle.min.
js"])
```

The preceding code will install Danfo.js and Plotly.js in Dnotebook. Danfo.js uses the installed Plotly.js to make plots. Plots won't work unless Plotly is explicitly loaded.

> **Note**
> Older versions of Danfo.js (pre-0.2.3) ship with Plotly.js. It has been removed in newer versions as stated in the release notes shown here: `https://danfo.jsdata.org/release-notes#latest-release-node-v-0-2-5-browser-0-2-4`.

If you're making plots in an HTML file, ensure to add the `script` tags to your header as demonstrated in the following code snippet:

```
...
<head>
<script src="https://cdn.plot.ly/plotly-1.2.0.min.js"></script>
<script src="https://cdn.jsdelivr.net/npm/danfojs@0.2.3/lib/bundle.min.js"></script>
</head>
...
```

Finally, in UI libraries such as React or Vue, ensure to install Danfo.js and Plotly.js via package managers such as npm or yarn.

# Downloading a dataset for plotting

In this section, we'll be downloading a real-world financial dataset, and this dataset will be used for all our examples. In Dnotebook, you can download the dataset once in a top cell and have it available to other cells as follows:

```
var financial_df;
dfd.read_csv("https://raw.githubusercontent.com/plotly/datasets/master/finance-charts-apple.csv")
    .then(data => {
        financial_df = data
    })
```

> **Note**
> Ensure to declare `financial_df` with `var`. This makes `financial_df` available to every cell in your Dnotebook. If working in React or plain HTML, then it is recommended to use `let` or `const`.

We can show the top five rows of `financial_df` using the `head` function and `table` as shown in the following code snippet:

```
table(financial_df.head())
```

Running the preceding code gives the following output:

|   | Date | AAPL.Open | AAPL.High | AAPL.Low | AAPL.Close | AAPL.Volume |
|---|------|-----------|-----------|----------|------------|-------------|
| 0 | 2015-02-17 | 127.489998 | 128.880005 | 126.919998 | 127.830002 | 63152400 |
| 1 | 2015-02-18 | 127.629997 | 128.779999 | 127.449997 | 128.720001 | 44891700 |
| 2 | 2015-02-19 | 128.479996 | 129.029999 | 128.330002 | 128.449997 | 37362400 |
| 3 | 2015-02-20 | 128.619995 | 129.5 | 128.050003 | 129.5 | 48948400 |
| 4 | 2015-02-23 | 130.020004 | 133 | 129.660004 | 133 | 70974100 |

Figure 6.1 – Table showing the top five rows of the financial dataset

Now that we have our datasets, we can start making some interesting plots. First, we'll start with a simple line chart.

# Creating line charts with Danfo.js

Line charts are simple chart types mostly used on Series data or single columns. They can show trends in data points. To make a line chart on a single column – say, `AAPL.Open` in `financial_df` – we can do the following:

```
var layout = {
    yaxis: {
       title: 'AAPL open points',
    }
}
var config = {
  displayModeBar: false,
  layout
}
financial_df ['AAPL.Open'].plot(this_div()).line(config)
```

Running the preceding code gives the following output:

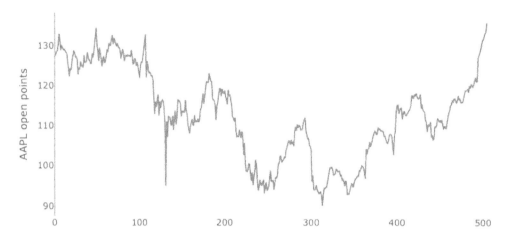

Figure 6.2 - Top five rows of the financial dataset

Notice that we are using DataFrame subsetting (`financial_df["column name"]`) to get a single column – `AAP1.Open` – as a Series. Then, we call the `.line` chart type and pass in a `config` object. The `config` object accepts the `layout` property as well as other arguments used by Danfo.js and Plotly.

If you want to plot specific columns, you can pass an array of column names to the `config` parameter, as demonstrated in the following code snippet:

```
var layout = {
    yaxis: {
        title: 'AAPL open points',
    }
}
var config = {
  columns: ["AAPL.Open", "AAPL.Close"],
  displayModeBar: true,

  layout
}
financial_df.plot(this_div()).line(config)
```

Running the preceding code gives the following output:

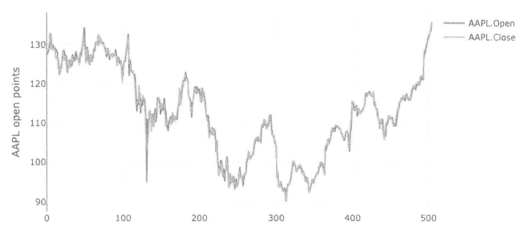

Figure 6.3 – Plotting two columns as a line chart

By default, the *x* axis of the chart is the index of the DataFrame or Series. In the case of the `financial_df` DataFrame, the index was auto-generated when we downloaded the dataset with the `read_csv` function. If you want to change the index, you can use the `set_index` function as shown in the following code snippet:

```
var new_df = financial_df.set_index({key: "Date"})
table(new_df.head())
```

The output is as follows:

| | AAPL.Open | AAPL.High | AAPL.Low | AAPL.Close | AAPL.Volume |
|---|---|---|---|---|---|
| **2015-02-17** | 127.489998 | 128.880005 | 126.919998 | 127.830002 | 63152400 |
| **2015-02-18** | 127.629997 | 128.779999 | 127.449997 | 128.720001 | 44891700 |
| **2015-02-19** | 128.479996 | 129.029999 | 128.330002 | 128.449997 | 37362400 |
| **2015-02-20** | 128.619995 | 129.5 | 128.050003 | 129.5 | 48948400 |
| **2015-02-23** | 130.020004 | 133 | 129.660004 | 133 | 70974100 |

Figure 6.4 – Table showing the top five rows with the index set to date

If we make the same plot as earlier, we see that the *x* axis is automatically formatted to dates:

Figure 6.5 – Chart of two columns (AAPL.open, AAPL.close) against the date index

You can specify other Plotly configuration options such as the width, fonts, and so on, by passing them to either the `layout` property or in the body of the `config` object. For example, to configure properties such as fonts, text size, layout width, and as far as even adding custom buttons to your plots, you can do the following:

```
var layout = {

...   legend: {bgcolor: "#fcba03",
            bordercolor: "#444",
            borderwidth: 1,
            font: {family: "Arial", size: 10, color: "#fff"}
    },
    ...}
var config = {
    columns: ["AAPL.Open", "AAPL.Close"],
    displayModeBar: true,
    modeBarButtonsToAdd: [{
        name: 'about',
        icon: Plotly.Icons.question,
        click: function (gd) {
          alert('An example of configuring Danfo.Js Plots')
        }
```

```
    }] ,
  layout
}
new_df.plot(this_div()).line(config)
```

Running the preceding code cell gives the following output:

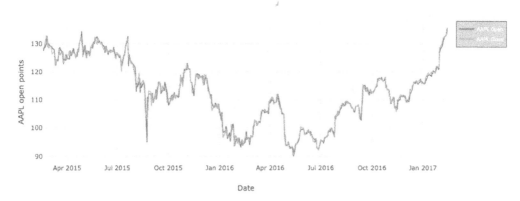

Figure 6.6 – A line plot with various configurations as well as layout properties specified

With the preceding information, you can get started making nice line charts from your dataset. In the next section, we'll cover another type of plot – a scatter plot – available in Danfo.js.

# Creating scatter plots with Danfo.js

We can easily make scatter plots by specifying the plot type to be scatter. For example, using the code from the preceding section, *Creating line charts with Danfo.js*, we can just change the plot type from line to scatter, and we get a scatter plot of the selected columns, as demonstrated in the following code block:

```
var layout = {
  title: "Time series plot of AAPL open and close points",
  width: 1000,
  yaxis: {
    title: 'AAPL open points',
  },
  xaxis: {
    title: 'Date',
```

```
  }
 }
 var config = {
   columns: ["AAPL.Open", "AAPL.Close"],
   layout
 }
 new_df.plot(this_div()).scatter(config)
```

Running the preceding code cell gives the following output:

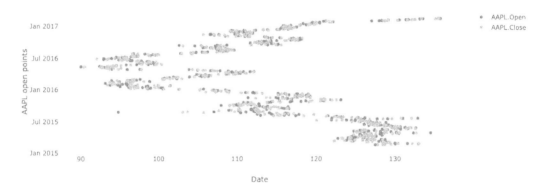

Figure 6.7 – A scatter plot on two columns

If you need to make a scatter plot between two specific columns in the DataFrame, you can specify x and y values in the config object, as shown in the following code:

```
 var layout = {
   title: "Time series plot of AAPL open and close points",
   width: 1000,
   yaxis: {
     title: 'AAPL open points',
   },
   xaxis: {
     title: 'Date',
   }
 }
 var config = {
   x: "AAPL.Low",
```

```
    y: "AAPL.High",
    layout
}
new_df.plot(this_div()).scatter(config)
```

Running the preceding code cell gives the following output:

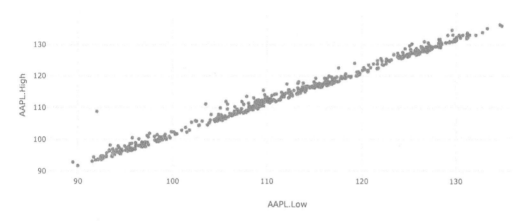

Figure 6.8 – A scatter plot on explicitly specified x and y columns

To customize the layout or set `config`, you can pass the corresponding options to the `config` object just like we did in the *Creating line charts with Danfo.js* section.

In the next section, we'll briefly look at two similar types of plots – box and violin plots.

# Creating box and violin plots with Danfo.js

The box and violin plots are similar and will generally use the same parameters. So, we will cover them both in this section.

In the following examples, we will first make a box plot and then change it to a violin plot by changing only the plot type option.

## Making box and violin plots for a Series

To make a box plot for a Series or a single column in a DataFrame, first, we subset to get the Series, and then we'll call the plot type on it, as demonstrated in the following code snippet:

```
var layout = {
    title: "Box plot on a Series",
}
var config = {
    layout
}
new_df["AAPL.Open"].plot(this_div()).box(config)
```

Running the preceding code cell gives the following output:

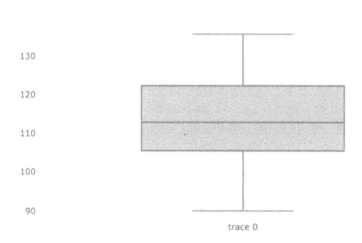

Figure 6.9 – A box plot on a Series

Now, in order to change the preceding plot to a violin plot, you can just change the plot type to `violin`, as shown in the following snippet:

```
. . .
new_df["AAPL.Open"].plot(this_div()).violin(config)
...
```

Running the preceding code cell gives the following output:

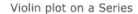

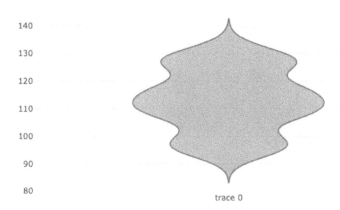

Figure 6.10 – A violin plot on a Series

What about when we need to make a box plot for more than one column at a time? Well, in the next section, we'll show you how.

## Box and violin plots for multiple columns

In order to make box/violin plots for more than one column in a DataFrame, you can pass an array of column names to the plot, as we demonstrate in the following code snippet:

```
var layout = {
    title: "Box plot on multiple columns",
}
var config = {
    columns: ["AAPL.Open", "AAPL.Close", "AAPL.Low", "AAPL.
High"],
    layout
}
new_df.plot(this_div()).box(config)
```

Running the preceding code cell gives the following output:

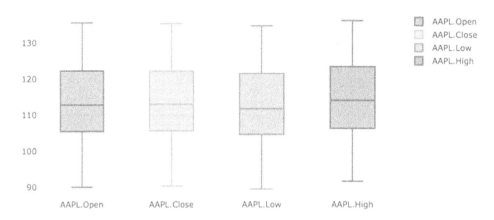

Figure 6.11 – Box plots of multiple columns at a time

Reusing the previous code, we can easily change the box plot to a violin plot by changing the plot type, as follows:

```
...
new_df.plot(this_div()).violin(config)
...
```

We get the following output:

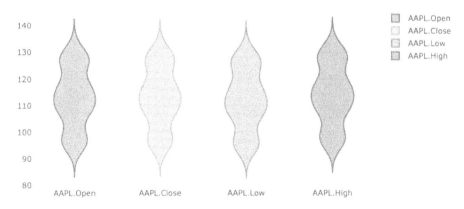

Figure 6.12 – Violin plots of multiple columns at a time

Finally, what if we want to specify the x and y values? We will show this in the following section.

## Box and violin plots with specific x and y values

We can make box and violin plots with specific x and y values. The x and y values are column names that must be present in the DataFrame.

> **Note**
> It is recommended that the x value of a box plot is categorical, that is, has a fixed number of classes. This ensures interpretability.

In the following example, we show you how to explicitly specify x and y values to a plot:

```
var layout = {
    title: "Box plot on x and y values",
}
var config = {
    x: "direction",
    y: "AAPL.Open",
    layout
}
new_df.plot(this_div()).box(config)
```

Running the preceding code cell gives the following output:

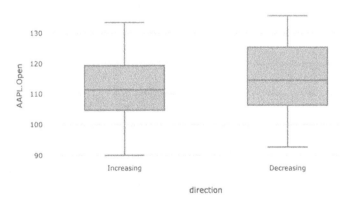

Figure 6.13 – Box plots from specific x and y values

Notice that the x value is a categorical variable called direction. This column has two fixed classes – Increasing and Decreasing.

As usual, we can get the corresponding violin plot by changing the type:

```
. . .
new_df.plot(this_div()).violin(config)
...
```

Running the preceding code cell gives the following output:

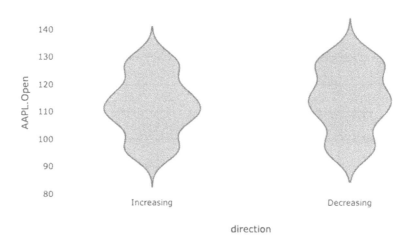

Figure 6.14 – Violin plot from specific x and y values

Now, what would happen if we specified both continuous values for x and y? Well, let's find out in the following example:

```
var layout = {
    title: "Box plot on two continuous variables",
}
var config = {
    x: "AAPL.Low",
    y: "AAPL.Open",
    layout
}
new_df.plot(this_div()).box(config)
```

Running the preceding code cell gives the following output:

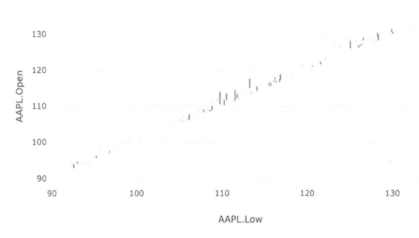

Figure 6.15 – Box plot on two continuous variables

From the preceding output, you can see that the chart becomes almost uninterpretable and the goal of using a box/violin plot is not achieved. So, therefore, it is recommended to use box/violin plots for categorical x values.

In the next section, we'll introduce the `hist` plot type for making histograms.

# Creating histograms with Danfo.js

A histogram, as we explained earlier, is a representation of the spread of data. The `hist` function exposed by the plot namespace can be used to make histograms from DataFrames or Series, as we'll demonstrate in the following section.

## Creating a histogram from a Series

In order to create a histogram from a Series, you can call the `hist` function on the Series, or if plotting on a DataFrame, you can subset the DataFrame with the column name, as shown in the following example:

```
var layout = {
    title: "Histogram on a Series data",
}
var config = {
```

```
    layout
}
new_df["AAPL.Open"].plot(this_div()).hist(config)
```

Running the preceding code cell gives the following output:

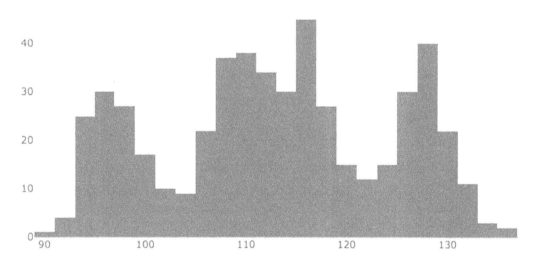

Figure 6.16 – Histogram on Series data

Next, we'll make a histogram for more than one column in a DataFrame at a time.

## Creating a histogram from multiple columns

If you want to make a histogram for more than one column in a DataFrame, you can pass the names of the columns as an array of column names, as we show in the following code snippet:

```
var layout = {
    title: "Histogram of two columns",
}
var config = {
    columns: ["dn", "AAPL.Adjusted"],
```

```
    layout
}
new_df.plot(this_div()).hist(config)
```

Running the preceding code cell gives the following output:

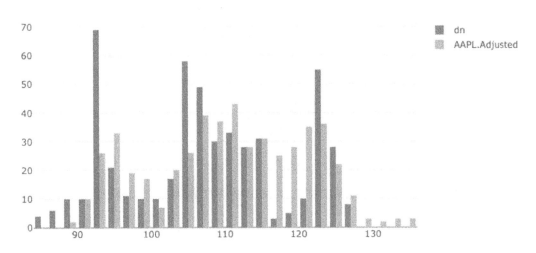

Figure 6.17 – Histogram on two columns

If you need to specify the single value x or y from which to generate the histogram, you can pass the x or y value to the `config` object.

> **Note**
>
> You can only specify one of x or y at a time. This is because a histogram is a univariate chart. So, if you specify x values, the histogram will be a vertical one, and if y is specified, it will be a horizontal histogram.

In the following example, we make a horizontal histogram by specifying y values:

```
var layout = {
    title: "A horizontal histogram",
}
var config = {
    y: "dn",
    layout
```

```
    }
new_df.plot(this_div()).hist(config)
```

Running the preceding code cell gives the following output:

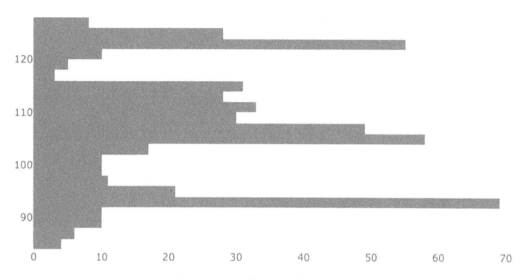

Figure 6.18 – A horizontal histogram

By default, histograms are vertical and that is equivalent to setting the x parameter.

In the next section, we'll introduce bar charts.

## Creating bar charts with Danfo.js

A bar chart presents categorical data with rectangular bars where the lengths are proportional to the values that they represent.

The `bar` function can also be called on the `plot` namespace and various configuration options can also be applied. In the following sections, we'll demonstrate how to create bar charts from a Series as well as a DataFrame with multiple columns.

# Creating a bar chart from a Series

To make a simple bar chart from a Series, you can do the following:

```
var layout = {
    title: "A simple bar chart on a series",
}
var config = {
   layout
}
new_df["AAPL.Volume"].plot(this_div()).bar(config)
```

Running the preceding code cell gives the following output:

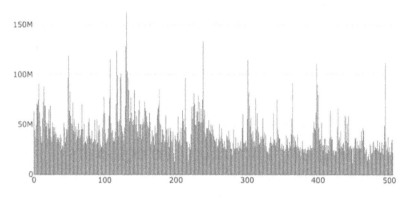

Figure 6.19 – A bar chart on a Series

Looking at the preceding figure, you'll notice that we have a large number of bars. This is because the AAPL.Volume column is a continuous variable and a bar is created for each point.

To avoid such uninterpretable charts, it is recommended to use bar charts for variables with a fixed number of numeric classes. We can demonstrate this by creating a simple Series, as shown in the following code:

```
custom_sf = new dfd.Series([1, 3, 2, 6, 10, 34, 40, 51, 90,
75])
custom_sf.plot(this_div()).bar(config)
```

Running the preceding code cell gives the following output:

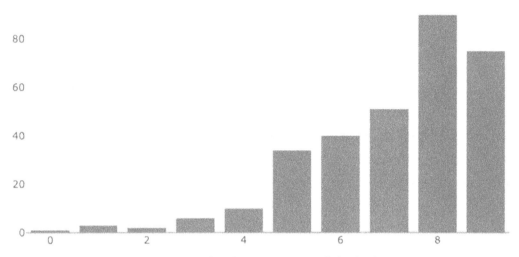

Figure 6.20 – A bar chart on a Series with fixed values

In the next section, we'll show you how to make grouped bar charts from a list of specified column names.

## Creating a bar chart from multiple columns

To create grouped bar charts from a list of column names, you can pass the names of the columns to the config object, as we demonstrate in the following example:

```
var layout = {
    title: "A bar chart on two columns",
}
var config = {
    columns: ["price", "cost"],
    layout
}
var df = new dfd.DataFrame({'price': [20, 18, 489, 675, 1776],
                           'cost': [40, 22, 21, 60, 19],
                           'count': [4, 25, 281, 600, 1900]},
                          {index: [1990, 1997, 2003, 2009,
```

```
2014]})
df.plot(this_div()).bar(config)
```

Running the preceding code cell gives the following output:

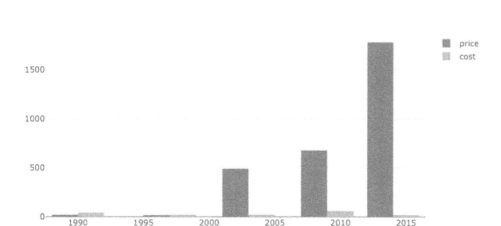

Figure 6.21 – A bar chart on two columns

Notice that in the preceding example, we created a new DataFrame. This is because the financial dataset does not contain the required data types for bar charts as we said earlier.

That brings us to the end of this chapter. Congratulations on making it this far! We are sure you have learned a lot and can use the knowledge gained here in your personal projects.

# Summary

In this chapter, we covered plotting and visualization with Danfo.js. First, we showed you how to set up Danfo.js and Plotly in a new project, and then moved on to downloading a dataset, which we loaded into a DataFrame. Next, we showed you how to create basic charts such as line, bar, and scatter plots, and then statistical charts such as histograms and box and violin plots. Finally, we showed you how to configure plots created with Danfo.js.

The knowledge you have gained in this and *Chapter 5, Data Visualization with Plotly.js*, will be of practical use when creating data-driven apps as well as custom dashboards.

In the next chapter, you'll learn about data aggregation and group-by operations, thereby understanding how to perform data transformations such as merging, joining, and concatenation.

# 7
# Data Aggregation and Group Operations

**Data aggregation** and **group operations** are very important methods in data analysis. These methods provide the ability to split data into a set of groups based on the specified key, and then apply some set of groupby operations (aggregations or transformations) to the grouped data to produce a new set of values. The resulting values are then combined into a single data group.

This approach is popularly known as **split-apply-combine**. The term was actually coined by Hadley Wickham, the author of many popular **R** packages, to describe group operations. *Figure 7.1* describes the idea of split-apply-combine graphically:

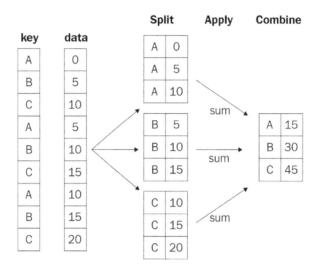

Figure 7.1 – groupby illustration

In this chapter, we look into ways of performing group operations: how to group data by column keys and perform data aggregation on grouped data jointly or independently.

This chapter will also show how to access grouped data by keys. It also gives insight into how to create a custom aggregate function for your data.

The following are the topics to be covered in this chapter:

- Grouping data
- Iterating through grouped data
- Using the .apply method
- Data aggregation of grouped data

# Technical requirements

In order to follow along with this chapter, you should have the following:

- A modern browser such as Chrome, Safari, Opera, or Firefox
- **Node.js**, **Danfo.js**, and **Dnotebook** installed on your system

The code for this chapter is available here: `https://github.com/PacktPublishing/Building-Data-Driven-Applications-with-Danfo.js/tree/main/Chapter07`.

# Grouping data

Danfo.js only provides the ability to group data by means of values in a specific column. For the current version of Danfo.js, the specified number of columns for grouping can only be one or two.

In this section, we will show how to group by single and double columns.

## Single-column grouping

First, let's start by creating a DataFrame and then group it by a single column:

```
let data = { 'A': [ 'foo', 'bar', 'foo', 'bar',
                    'foo', 'bar', 'foo', 'foo' ],
                'C': [ 1, 3, 2, 4, 5, 2, 6, 7 ],
            'D': [ 3, 2, 4, 1, 5, 6, 7, 8 ] };
let df = new dfd.DataFrame(data);
let group_df = df.groupby([ "A"]);
```

The preceding code involves the following steps:

1. First, we create a DataFrame by using an `object` method.
2. We then make a call to the `groupby` method.
3. Then, we specify that the DataFrame should be grouped by column A.

`df.groupby(['A'])` returns a `groupby` data structure that contains all the necessary methods needed for grouping data.

We can decide to perform our data operations on all the columns grouped by A or we specify any other column.

The preceding code outputs the following table:

| | A | C | D |
|---|---|---|---|
| 0 | foo | 1 | 3 |
| 1 | bar | 3 | 2 |
| 2 | foo | 2 | 4 |
| 3 | bar | 4 | 1 |
| 4 | foo | 5 | 5 |
| 5 | bar | 2 | 6 |
| 6 | foo | 6 | 7 |
| 7 | foo | 7 | 8 |

Figure 7.2 – DataFrame

In the following code, we will see how we can perform some common `groupby` operations on the grouped data:

```
group_df.mean().print()
```

Using the `groupby_df` operations created in the preceding code snippet, we make a call to the `groupby mean` method. This method calculates the mean per group as shown in the following figure:

| | A | C_mean | D_mean |
|---|---|---|---|
| 0 | foo | 4.19999980926... | 5.40000009536... |
| 1 | bar | 3 | 3 |

Figure 7.3 – groupby DataFrame

The following figure shows the operation of the preceding code graphically and how the preceding table output is generated:

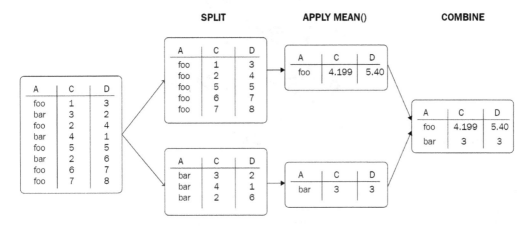

Figure 7.4 – Graphical depiction of the groupby method

Based on the split-apply-combine method that we discussed at the beginning of this chapter, df.groupby(['A']) groups the DataFrame into two keys – foo and bar.

With the DataFrame grouped into foo and bar keys, the values in other columns (C and D) are assigned to each of these keys respectively based on their row alignment. To nail this point, if we pick the bar key, from *Figure 7.2*, we can see that column C has three data points (3, 4, 2) in the bar rows.

Hence, if we were to perform a data aggregation operation, such as calculating the mean of the data assigned to the bar key, the data point belonging to column C assigned to the bar key will have a mean of 3, which corresponds to the table in *Figure 7.3*.

---

**Note**

The same operation as described in the preceding paragraph occurs for the foo key and all other data points are assigned

---

As I said before, this call to the group mean method is applied to all the columns grouped by A. Let's choose a column to which we would like to specifically apply the group operation, as follows:

```
let col_c = group_df.col(['C'])
col_c.sum().print()
```

First, we make a call to the `col` method in `groupby`. This method takes in an array of column names. The main purpose of the preceding code is to obtain the sum of column C for each of the grouped keys (`foo` and `bar`) and this gives the following table output:

| | A | C_sum |
|---|---|---|
| 0 | foo | 21 |
| 1 | bar | 9 |

Figure 7.5 – The groupby operation on column C

The same operation can be applied to any other column in the grouped data, as follows:

```
let col_d = group_df.col(['D'])
col_d.count().print()
```

This code snippet follows the same approach as the preceding code, just that the `count` `groupby` operation is applied to column D, which gives us the following output:

| | A | D_count |
|---|---|---|
| 0 | foo | 5 |
| 1 | bar | 3 |

Figure 7.6 – The groupby "count" operation on column D

A DataFrame can also be grouped by two columns; the operation shown for single-column grouping is also applicable to two-column grouping, as we will see in the next sub-section.

## Double-column grouping

First, let's create a DataFrame and add an extra column as follows:

```
let data = { 'A': [ 'foo', 'bar', 'foo', 'bar',
                    'foo', 'bar', 'foo', 'foo' ],
             'B': [ 'one', 'one', 'two', 'three',
                    'two', 'two', 'one', 'three' ],
             'C': [ 1, 3, 2, 4, 5, 2, 6, 7 ],
             'D': [ 3, 2, 4, 1, 5, 6, 7, 8 ] };
let df = new dfd.DataFrame(data);
let group_df = df.groupby([ "A", "B"]);
```

We added an extra column, B, which contains categoric data – one, two, and three. The DataFrame is grouped by columns A and B, as shown in the following table:

| | A | B | C | D |
|---|---|---|---|---|
| 0 | foo | one | 1 | 3 |
| 1 | bar | one | 3 | 2 |
| 2 | foo | two | 2 | 4 |
| 3 | bar | three | 4 | 1 |
| 4 | foo | two | 5 | 5 |
| 5 | bar | two | 2 | 6 |
| 6 | foo | one | 6 | 7 |
| 7 | foo | three | 7 | 8 |

Figure 7.7 – DataFrame including column B

We can also calculate the mean as shown with the single-column grouping, as follows:

```
group_df.mean().print()
```

This applies the mean to columns C and D based on the group key of columns A and B. For example, in *Figure 7.7*, we can see that we have a group key from columns A and B named (foo, one). This key occurs twice in the data.

The preceding code outputs the following table:

| | A | B | C_mean | D_mean |
|---|---|---|---|---|
| 0 | foo | one | 3.5 | 5 |
| 1 | foo | two | 3.5 | 4.5 |
| 2 | foo | three | 7 | 8 |
| 3 | bar | one | 3 | 2 |
| 4 | bar | two | 2 | 6 |
| 5 | bar | three | 4 | 1 |

Figure 7.8 – The groupby mean of the groupby A and B DataFrame

In *Figure 7.7*, column C has the values 1 and 6 belonging to the (foo, one) key. Also, D has the values 3 and 7 belonging to the same key. If we were to take the mean, we would see that it corresponds to the first column in *Figure 7.8*.

We can also go ahead and obtain the sum for one of the columns grouped by columns A and B. Let's choose column C:

```
let col_c = group_df.col(['C']);
col_c.sum().print();
```

We obtained column C from the grouped data and then calculated the sum per the group keys, as shown in the following table:

|   | A | B | C_sum |
|---|---|---|-------|
| 0 | foo | one | 7 |
| 1 | foo | two | 7 |
| 2 | foo | three | 7 |
| 3 | bar | one | 3 |
| 4 | bar | two | 2 |
| 5 | bar | three | 4 |

Figure 7.9 – Sum of column C per group

In this section, we saw how we can group data by single or double columns. We also looked into performing data aggregation and accessing column data of a grouped DataFrame. In the next section, we'll look into how to access grouped data per grouped keys.

# Iterating through grouped data

In this section, we will see how to access grouped data based on grouped keys, loop through this grouped data, and perform data aggregation operations on it.

## Iterating through single- and double-column grouped data

In this section, we will see how Danfo.js provides the means of iterating through each of the groups created during the groupby operations. This data is grouped by the keys contained in the groupby column.

The keys are stored as a dictionary or object in a class attribute called `data_tensors`. This object contains the grouped key as its keys and also stores the DataFrame data associated with the keys as the object values.

Using the previous DataFrame, let's group by column A and then iterate through `data_tensors`:

```
let group_df = df.groupby([ "A"]);
console.log(group_df.data_tensors)
```

We group the DataFrame by column A and then print out `data_tensors`, as shown in the following screenshot:

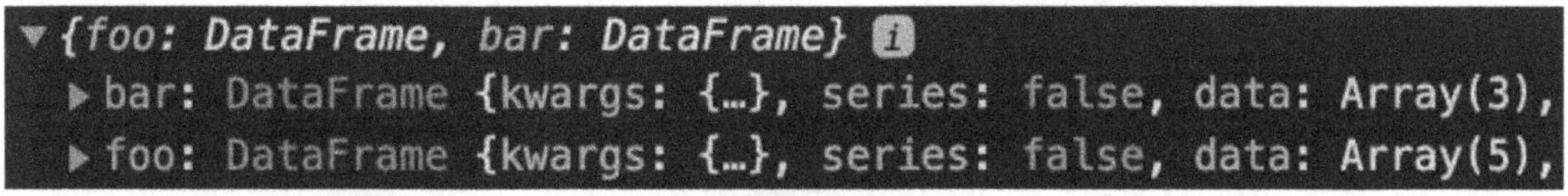

Figure 7.10 – data_tensors output

*Figure 7.10* contains more detailed information about what the `data_tensors` attribute contains, but the whole content can be summarized into the following structure:

```
{
    foo: DataFrame,
    bar: DataFrame
}
```

The keys are the values of column A and the values for the keys are the DataFrame associated with these keys.

We can iterate through `data_tensors` and print out the `DataFrame` table to see what they contain, as shown in the following code:

```
let grouped_data = group_df.data_tensors;
for (let key in grouped_data) {
    grouped_data[key].print();
}
```

First, we access `data_tensors`, which is a `groupby` class attribute, and assign it to a variable called `grouped_data`. We then loop through `grouped_data`, access each of its keys, and print their corresponding DataFrame as a table, as shown in the following screenshot:

| | A | B | C | D |
|---|---|---|---|---|
| 0 | foo | one | 1 | 3 |
| 1 | foo | two | 2 | 4 |
| 2 | foo | two | 5 | 5 |
| 3 | foo | one | 6 | 7 |
| 4 | foo | three | 7 | 8 |

| | A | B | C | D |
|---|---|---|---|---|
| 0 | bar | one | 3 | 2 |
| 1 | bar | three | 4 | 1 |
| 2 | bar | two | 2 | 6 |

Figure 7.11 – groupby keys and their DataFrame

Also, we can apply this same approach, as shown in the preceding code, to data grouped by two columns:

```
let group_df = df.groupby([ "A", "B"]);
let grouped_data = group_df.data_tensors;
for (let key in grouped_data) {
  let key_data = grouped_data[key];
  for (let key2 in key_data) {
    grouped_data[key][key2].print();
  }
}
```

Here are the steps we followed in the preceding code snippet:

1. First of all, the `df` DataFrame is grouped by two columns (A and B).

2. We assign `data_tensors` to a variable called `grouped_data`.

3. We loop through `grouped_data` to obtain the keys.

4. We loop through the `grouped_data` object and also loop through its inner object (`key_data`) per key, due to the object data format generated for `grouped_data`, as shown:

```
{
    foo : {
        one: DataFrame,
        two: DataFrame,
        three: DataFrame
    },
    bar: {
        one: DataFrame,
        two: DataFrame,
        three: DataFrame
    }
}
```

The code snippet gives us the following output:

|   | A | B | C | D |
|---|---|---|---|---|
| 0 | foo | one | 1 | 3 |
| 1 | foo | one | 6 | 7 |

|   | A | B | C | D |
|---|---|---|---|---|
| 0 | foo | two | 2 | 4 |
| 1 | foo | two | 5 | 5 |

|   | A | B | C | D |
|---|---|---|---|---|
| 0 | foo | three | 7 | 8 |

|   | A | B | C | D |
|---|---|---|---|---|
| 0 | bar | one | 3 | 2 |

|   | A | B | C | D |
|---|---|---|---|---|
| 0 | bar | two | 2 | 6 |

|   | A | B | C | D |
|---|---|---|---|---|
| 0 | bar | three | 4 | 1 |

Figure 7.12 – The output of a two-column grouping of a DataFrame

In this section, we went through how to iterate through grouped data. We saw how the object format for `data_tensor` varies by how the data is being grouped, either by single or double columns.

We saw how to iterate `data_tensor` to obtain the keys and their associated data. In the next sub-section, we will see how we can obtain the data associated with each key without looping through `data_tensor` manually.

## Using the get_groups method

Danfo.js provides a method called `get_groups()` that enables easy access to each of the key-value DataFrames without looping through the `data_tensors` object. This is handy whenever we need to access particular data belonging to a set of key combinations.

Let's start with a single-column grouping, as follows:

```
let group_df = df.groupby([ "A"]);
group_df.get_groups(["foo"]).print()
```

We group by column A, and then make a call to the `get_groups` method. The `get_groups` method takes in a key combination as an array.

For a single-column grouping, we only have a single key combination. Hence, we pass in one of the keys called `foo` and then print out the corresponding grouped data, as shown in the following screenshot:

|   | A | B | C | D |
|---|---|---|---|---|
| 0 | foo | one | 1 | 3 |
| 1 | foo | two | 2 | 4 |
| 2 | foo | two | 5 | 5 |
| 3 | foo | one | 6 | 7 |
| 4 | foo | three | 7 | 8 |

Figure 7.13 – get_groups of the foo key

Also, the same thing can be applied to all other keys as follows:

```
group_df.get_groups(["bar"]).print()
```

The preceding code gives us the following output:

| | A | B | C | D |
|---|---|---|---|---|
| 0 | bar | one | 3 | 2 |
| 1 | bar | three | 4 | 1 |
| 2 | bar | two | 2 | 6 |

Figure 7.14 – get_groups for the bar key

The same approach as shown for the single-column, groupby is also applicable to two-column grouping:

```
let group_df = df.groupby([ "A", "B"]);
group_df.get_groups(["foo","one"]).print()
```

Remember that the get_groups method takes in the combination of keys as an array. Hence, for two-column grouping, we pass in the key combination of columns A and B that we want. Therefore, we obtain the following output:

| | A | B | C | D |
|---|---|---|---|---|
| 0 | foo | one | 1 | 3 |
| 1 | foo | one | 6 | 7 |

Figure 7.15 – Obtaining a DataFrame for the foo key and one combination

The same thing can be done for any other key combination. Let's try for the bar and two keys, as follows:

```
group_df.get_groups(["bar","two"]).print()
```

We obtain the following output:

| | A | B | C | D |
|---|---|---|---|---|
| 0 | bar | two | 2 | 6 |

Figure 7.16 – DataFrame for the bar and two keys

In this section, we went through how to iterate grouped data, data grouped by either single or double columns. We also saw how the internal `data_tensor` data object is formatted based on how the data is being grouped. We also saw how to access data associated with each grouped key without looping.

In the next section, we'll also look into creating custom data aggregation functions by using the `.apply` method.

# Using the .apply method

In this section, we will be using the `.apply` method to create custom data aggregation functions that can be applied to our grouped data.

The `.apply` method enables custom functions to be applied to grouped data. It is the major function of the split-apply-combine method discussed earlier in this chapter.

The `groupby` method implemented in Danfo.js only contains a small set of data aggregation methods needed for group data, hence the `.apply` method gives users the ability to construct a special data aggregation method from the grouped data.

Using the previous data, we will create a new DataFrame excluding column B as seen in the previous DataFrame, and then create a custom function that will be applied to the grouped data:

```
let group_df = df.groupby([ "A"]);
const add = (x) => {
   return x.add(2);
};
group_df.apply(add).print();
```

In the preceding code, we group the DataFrame by column A, and then proceed to create a custom function called `add` that adds the value 2 to all the data points in the grouped data.

> **Note**
>
> The parameter to be passed into this function, `add`, and similar functions, such as `sub`, `mul`, and `div`, can be either a DataFrame, Series, array, or an integer value.

The preceding code generates the following output:

| | A | C_apply | D_apply |
|---|---|---|---|
| 0 | foo | 3 | 5 |
| 1 | foo | 4 | 6 |
| 2 | foo | 7 | 7 |
| 3 | foo | 8 | 9 |
| 4 | foo | 9 | 10 |
| 5 | bar | 5 | 4 |
| 6 | bar | 6 | 3 |
| 7 | bar | 4 | 8 |

Figure 7.17 – Applying a custom function to group data

Let's create another custom function to subtract the minimum value of each grouped data from each of the group values:

```
let data = { 'A': [ 'foo', 'bar', 'foo', 'bar',
                    'foo', 'bar', 'foo', 'foo' ],
             'B': [ 'one', 'one', 'two', 'three',
                    'two', 'two', 'one', 'three' ],
             'C': [ 1, 3, 2, 4, 5, 2, 6, 7 ],
             'D': [ 3, 2, 4, 1, 5, 6, 7, 8 ] };
let df = new DataFrame(data);
let group_df = df.groupby([ "A", "B"]);

const subMin = (x) => {
  return x.sub(x.min());
};


group_df.apply(subMin).print();
```

First, we created a DataFrame containing columns A, B, C, and D. Then, we grouped the DataFrame by columns A and C.

A custom function called subMin is created to take in grouped data, to obtain the minimum values of the grouped data, and to subtract the minimum value from each of the data points in the grouped data.

This custom function is then applied to the group_df grouped data via the .apply method, hence we obtain the following output table:

| | A | B | C_apply | D_apply |
|---|---|---|---|---|
| 0 | foo | one | 0 | 0 |
| 1 | foo | one | 5 | 4 |
| 2 | foo | two | 0 | 0 |
| 3 | foo | two | 3 | 1 |
| 4 | foo | three | 0 | 0 |
| 5 | bar | one | 0 | 0 |
| 6 | bar | two | 0 | 0 |
| 7 | bar | three | 0 | 0 |

Figure 7.18 – The subMin custom apply function

If we look at the table in the preceding figure, we can see that some group data only occurs once, such as group data belonging to the bar and two, bar and one, bar and three, and foo and three keys.

The group data belonging to previous key has only one item, hence the minimum is also the single value contained in the group; therefore, it has a value of 0 for the C_apply and D_apply columns.

We can adjust the subMin custom function to only subtract the minimum from each value if the key pair has more than one row, as follows:

```
const subMin = (x) => {
  if (x.values.length > 1) {
    return x.sub(x.min());
  } else {
    return x;
  }
};


group_df.apply(subMin).print();
```

The custom function gives us the following output table:

| | A | B | C_apply | D_apply |
|---|---|---|---|---|
| 0 | foo | one | 0 | 0 |
| 1 | foo | one | 5 | 4 |
| 2 | foo | two | 0 | 0 |
| 3 | foo | two | 3 | 1 |
| 4 | foo | three | 7 | 8 |
| 5 | bar | one | 3 | 2 |
| 6 | bar | two | 2 | 6 |
| 7 | bar | three | 4 | 1 |

Figure 7.19 – Custom apply function

The following figure shows a graphical representation of the preceding code:

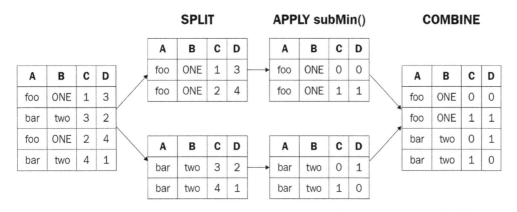

Figure 7.20 – groupby and subMin apply method illustration

The .apply method also gives us the ability to perform a data normalization process per group on the data.

In machine learning, we have what we called **standardization**, which involves rescaling the data between the range -1 and 1. This process of standardization involves subtracting the mean of the data from the data and then dividing by the standard deviation.

Using the preceding DataFrame, let's create a custom function to apply standardization to the data per group:

```
let data = { 'A': [ 'foo', 'bar', 'foo', 'bar',
                    'foo', 'bar', 'foo', 'foo' ],
             'C': [ 1, 3, 2, 4, 5, 2, 6, 7 ],
             'D': [ 3, 2, 4, 1, 5, 6, 7, 8 ] };
let df = new DataFrame(data);
let group_df = df.groupby([ "A"]);

// (x - x.mean()) / x.std()
const norm = (x) => {
  return x.sub(x.mean()).div(x.std());
};

group_df.apply(norm).print();
```

In the preceding code, we first group the data by column A. We then create a custom function called norm, which contains the standardization process that is being applied to the data, to give the following output:

| | A | C_apply | D_apply |
|---|------|-----------------|-----------------|
| 0 | foo | −1.2362678050... | −1.1573828458... |
| 1 | foo | −0.8499340415... | −0.6751399636... |
| 2 | foo | 0.30906704068... | −0.1928971707... |
| 3 | foo | 0.69540071487... | 0.77158844470... |
| 4 | foo | 1.08173441886... | 1.25383126735... |
| 5 | bar | 0 | −0.3779644966... |
| 6 | bar | 1 | −0.7559289932... |
| 7 | bar | −1 | 1.13389348983... |

Figure 7.21 – Standardizing grouped data

We've seen how to use the `.apply` method to create custom functions for `groupby` operations. Hence, we can create custom functions as needed based on the kind of operation we want to perform on the data.

In the next section, we'll look into data aggregation and how to assign different data aggregation operations to different columns of grouped data.

# Data aggregation of grouped data

Data aggregation involves the process of gathering data and presenting it in a summary form, such as showing its statistics. Aggregation itself is the process of gathering data for statistical purposes and presenting it as a number. In this section, we look at how to perform data aggregation in Danfo.js

Here is a list of all the available aggregation methods:

- `mean()`: To calculate the mean of grouped data
- `std()`: To calculate the standard deviation
- `sum()`: To get the sum of values in a group
- `count()`: To count the total number of values per group
- `min()`: To get the minimum value per group
- `max()`: To get the maximum value per group

At the start of this chapter, we saw how we can call some of the aggregate methods listed previously on group data. The `groupby` class also contains a method called `.agg`, which allows us to apply different aggregate operations to different columns at the same time.

## Data aggregation on single-column grouping

We will create a DataFrame and group the DataFrame by a column, and then apply two different aggregation methods on different columns:

```
let data = { 'A': [ 'foo', 'bar', 'foo', 'bar',
        'foo', 'bar', 'foo', 'foo' ],
    'C': [ 1, 3, 2, 4, 5, 2, 6, 7 ],
    'D': [ 3, 2, 4, 1, 5, 6, 7, 8 ] };
let df = new DataFrame(data);
```

```
let group_df = df.groupby([ "A"]);
```

```
group_df.agg({ C:"mean", D: "count" }).print();
```

We created a DataFrame and then grouped the DataFrame by column A. The grouped data is then aggregated by calling the .agg method.

The .agg method takes in an object whose keys are names of columns in the DataFrame and the values are the aggregation methods we want to apply to each of the columns. In the preceding block of code, we specified the keys to be C and D and the values to be mean and count:

| | A | C_mean | D_count |
|---|---|---|---|
| 0 | foo | 4.19999980926... | 5 |
| 1 | bar | 3 | 3 |

Figure 7.22 – Aggregation method on group data

We've seen how data aggregation can be done on single-column grouping. Let's now see how to perform the same operation on a DataFrame grouped by double columns.

## Data aggregation on double-column grouping

For double-column grouping, let's apply the same aggregation methods:

```
let data = { 'A': [ 'foo', 'bar', 'foo', 'bar',
        'foo', 'bar', 'foo', 'foo' ],
    'B': [ 'one', 'one', 'two', 'three',
        'two', 'two', 'one', 'three' ],
    'C': [ 1, 3, 2, 4, 5, 2, 6, 7 ],
    'D': [ 3, 2, 4, 1, 5, 6, 7, 8 ] };
let df = new DataFrame(data);
let group_df = df.groupby([ "A", "B"]);
```

```
group_df.agg({ C:"mean", D: "count" }).print();
```

The preceding code gives the following output:

| | A | B | C_mean | D_count |
|---|---|---|---|---|
| 0 | foo | one | 3.5 | 2 |
| 1 | foo | two | 3.5 | 2 |
| 2 | foo | three | 7 | 1 |
| 3 | bar | one | 3 | 1 |
| 4 | bar | two | 2 | 1 |
| 5 | bar | three | 4 | 1 |

Figure 7.23 – Aggregation method on two-column grouped data

In this section, we've seen how we can use the `.apply` method to create a custom function for grouped data, and also how to perform joint data aggregation on each column of the grouped data. The examples shown here can be extended to any specific data and custom functions can be created as desired.

## A simple application of groupby on real data

We've seen how to use `groupby` methods on dummy data. In this section, we will see how we can use `groupby` to analyze data.

We'll make use of the popular `titanic` dataset available here: `https://web.stanford.edu/class/archive/cs/cs109/cs109.1166/stuff/titanic.csv`. We will see how we can estimate the average number of people that survive the Titanic accident based on their gender and their class.

Let's read the `titanic` dataset into a DataFrame and output some of its rows:

```
const dfd = require('danfojs-node')
async function analysis(){
  const df = dfd.read_csv("titanic.csv")
  df.head().print()
}
analysis()
```

The preceding code should output the following table:

| | Survived | Pclass | Name | Sex | ... | Age |
|---|---|---|---|---|---|---|
| 0 | 0 | 3 | Mr. Owen Harr... | male | ... | 22 |
| 1 | 1 | 1 | Mrs. John Bra... | female | ... | 38 |
| 2 | 1 | 3 | Miss. Laina H... | female | ... | 26 |
| 3 | 1 | 1 | Mrs. Jacques ... | female | ... | 35 |
| 4 | 0 | 3 | Mr. William H... | male | ... | 35 |

Figure 7.24 – DataFrame table

From the dataset, we want to estimate the average number of people that survived based on their sex (the Sex column) and their traveling class (the Pclass column).

The following code shows how to estimate the average rate of survival as described previously:

```
const dfd = require('danfojs-node')
async function analysis(){
  const df = dfd.read_csv("titanic.csv")
  df.head().print()

  //groupby Sex column
  const sex_df = df.groupby(["Sex"]).col(["Survived"]).mean()
  sex_df.head().print()

  //groupby Pclass column
  const pclass_df = df.groupby(["Pclass"]).col(["Survived"]).
mean()
  pclass_df.head().print()
}
analysis()
```

The following table shows the average survival per sex:

| | Sex | Survived_mean |
|---|---|---|
| 0 | male | 0.19022688269... |
| 1 | female | 0.74203819036... |

Figure 7.25 – Average rate of survival based on sex

The following table shows the average survival per class:

| | Pclass | Survived_mean |
|---|---|---|
| 0 | 1 | 0.62962961196... |
| 1 | 2 | 0.47282609343... |
| 2 | 3 | 0.24435319006... |

Figure 7.26 – Average rate of survival based on Pclass

In this section, we saw a brief introduction of how to use the `groupby` operation to analyze real-life data.

# Summary

In this chapter, we extensively discussed the `groupby` operation as implemented in Danfo.js. We discussed grouping data and mentioned that at the moment, Danfo.js only supports grouping by single and double columns; there is a plan to make this more flexible in coming versions of Danfo.js. We also showed how to iterate through grouped data and access group keys and their associated grouped data. We looked at how to obtain grouped data associated with a group key without looping.

We also saw how the `.apply` method gives us the ability to create custom data aggregation functions for our grouped data, and finally, we demonstrated how to perform different aggregation functions on different columns of grouped data at the same time.

This chapter equipped us with the knowledge of grouping our data, and more essentially, it introduced us to the internals of Danfo.js. With this, we can reshape the `groupby` method to our desired taste and have the ability to contribute to Danfo.js.

In the next chapter, we will move on to more application basics, including how to use Danfo.js to build a data analysis web app, a no-code environment. We will also see how to turn Danfo.js methods into React components.

# Section 3: Building Data-Driven Applications

In *Section 3*, we switch to a hands-on approach and show you how to build data-driven applications. First, we show you how to build a no-code environment for data analysis and data handling. Then, we introduce you to the basics of machine learning with Danfo. js and TensorFlow.js. Next, we introduce you to recommendation systems and show you how to build one. Finally, we end this section by showing you how to create a Twitter analytics dashboard.

This section comprises the following chapters:

- *Chapter 8, Creating a No-Code Data Analysis/Handling System*
- *Chapter 9, Basics of Machine Learning*
- *Chapter 10, Introduction to TensorFlow.js*
- *Chapter 11, Building a Recommendation System with Danfo.js and TensorFlow.js*
- *Chapter 12, Building a Twitter Analysis Dashboard*
- *Chapter 13, Appendix: Essential JavaScript Concepts*

# 8
# Creating a No-Code Data Analysis/ Handling System

One of the main purposes of creating **Danfo.js** was to easily enable data processing in the browser. This gives the ability to integrate data analysis and handling data seamlessly into web apps. Apart from the ability to add data handling to a web app, we have the tools to make data handling and analysis look more like what designers do when they use **Photoshop** and **Figma**; how they mix brush strokes together on the canvas just with a click or how they manipulate images by laying a canvas on top the canvas, just by dragging and dropping and with some button clicks.

With Danfo.js, we can easily enable such an environment (using tools such as **React.js** and **Vue.js**) where data scientists become artists maneuvering their way through data with a few clicks of a button and getting the desired output without actually coding anything.

A lot of tools with such features commonly exist, but the cool thing about Danfo.js is building the whole app with tools in JavaScript. In fact, doing all operations in the browser without a call to the server is quite amazing.

The goal of this chapter is to show how such an environment can be built using Danfo.js and React.js. Also, note that the tools used here (apart from Danfo.js) are not mandatory for building the app; these are just the tools I'm quite familiar with.

This chapter will cover the following topics:

- Setting up the project environment
- Structuring and designing the app
- App layout and the `DataTable` component
- Creating different `DataFrame` operation components
- Implementing the `Chart` component

# Technical requirements

The following are the basic environment and knowledge requirements for this chapter:

- A modern web browser such as **Chrome**
- A suitable code editor such as **VScode**
- **Node.js** installed
- A bit of knowledge of `tailwindcss` and `React-chart-js`
- Knowledge of the basics of React.js is needed. To brush up on React.js, check out the official site at `https://reactjs.org/docs/hello-world.html`
- The code for this chapter is available and can be cloned from GitHub at `https://github.com/PacktPublishing/Building-Data-Driven-Applications-with-Danfo.js/tree/main/Chapter08`

# Setting up the project environment

React.js is used for the project, and to set up the React app, we will use the `create-react-app` package to automatically generate a frontend build pipeline for us. But first, make sure you have Node.js and **npm** installed in order to use `npx`, a package runner tool that comes with *npm 5.2+*.

Before we dive into setting up our environment, here are the tools needed and that will be installed in this chapter:

- **React.js**: A JavaScript framework for building the UI
- **Draggable**: A drag-and-drop library that makes it possible to move HTML elements around
- **React-chart-js**: A React library for `chart` components
- **React-table-v6**: A React library for displaying tables

The following are some alternatives to the preceding tools:

- **Vue.js**: A JavaScript library for building the UI
- **rechart.js**: A composable charting library built on React.js
- **Material-table**: A data table for React based on **material-UI**

To create a React app pipeline, we make a call to `create-react-app` using `npx` and then specify the name of our project as follows:

```
$ npx create-react-app data-art
```

This command will create a directory called `data-art` in the parent directory in which the command is being initiated. This `data-art` directory is prefilled with the React.js template and all the packages needed.

Here is the structure of the `data-art` folder:

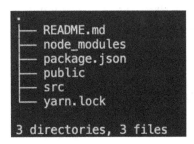

```
├── README.md
├── node_modules
├── package.json
├── public
├── src
└── yarn.lock

3 directories, 3 files
```

Figure 8.1 – React.js directory structure

After the installation, we can always start the app by using the following command (assuming you are not in the `data-art` directory in the terminal):

```
$ cd data-art
$ yarn start
```

The following command will start the app server and also output the server port in which the app is running in the terminal:

Figure 8.2 – yarn start output

As seen in *Figure 8.1*, the app is served at `http://localhost:3000`. If everything works properly, then once the server is started, it will automatically open the web browser to display the React app.

For the development of the app, we won't spend more time on styling, but we will make it easier to integrate styling in the future and also enable quick prototyping; we'll make use of `tailwindcss`.

To make Tailwind work with the React.js app, we will need to do some extra configuration.

Let's install Tailwind and its peer dependencies via `npm` as shown in the `tailwindcss` docs: `https://tailwindcss.com/docs/guides/create-react-app`:

```
npm install -D tailwindcss@npm:@tailwindcss/postcss7-compat
postcss@^7 autoprefixer@^9
```

Once the installation is done, we will go ahead and install the `craco` module, which allows us to override `postcss` configuration as follows:

```
npm install @craco/craco
```

After installing `craco`, we can go ahead and configure how to build, start, and test the React app. This will be done by changing the commands for `"start"`, `"build"`, and `"test"` in `package.json` to the following:

```
{
    "start": "craco start",
    "build": "craco build",
    "test": "craco test",
}
```

With the changes made previously, let's create a config file that enables `craco` to always inject `tailwindcss` and `autoprefixer` when building a React app, as shown in the following code:

```
// craco.config.js
module.exports = {
  style: {
    postcss: {
      plugins: [
        require('tailwindcss'),
        require('autoprefixer'),
      ],
    },
  },
}
```

Let's configure `tailwindcss` itself. With this configuration, we can tell `tailwindcss` to remove unused styles in production, we can add custom themes, and we can also add customized color, font, width, and height not included in the `tailwindcss` package, as follows:

```
//tailwind.config.js
module.exports = {
  purge: ["./src/**/*.{js,jsx,ts,tsx}", "./public/index.html"],
  darkMode: false, // or 'media' or 'class'
  theme: {
    extend: {},
  },
  variants: {
    extend: {},
  },
  plugins: [],
};
```

After configuring Tailwind, we will edit the `css` file, `index.css`, in the `src` directory. We'll add the following to the file:

```
/* ./src/index.css */
@tailwind base;
```

```
@tailwind components;
@tailwind utilities;
```

We are done with the configuration; we can now import `index.css` in `index.js`:

```
//index.js
. . . . . .
import "./index.css
. . . . . .
```

Note that in `App.js`, we still have the default code that came with the `create-react-app` package; let's edit the code. Here is the initial code:

```
function App() {
  return (
    <div className="App">
      <header className="App-header">
        <img src={logo} className="App-logo" alt="logo" />
        <p>
          Edit <code>src/App.js</code> and save to reload.
        </p>
        <a
          className="App-link"
          href="https://reactjs.org"
          target="_blank"
          rel="noopener noreferrer"
        >
          Learn React
        </a>
      </header>
    </div>
  );
}
```

We edit the HTML code in the `App` comment by editing the HTML and replacing it with the name of the app:

```
function App() {
  return (
```

```
        <div className="">
            Data-Art
        </div>
    );
}
```

By updating App.js with the preceding code and saving it, you should see the changes being made in the browser directly, as seen in the following screenshot:

Data-Art

Figure 8.3 – React app

Let's test our tailwindcss configuration to ensure it is set properly. We will do this by adding some style to the preceding code as follows:

```
function App() {
    return (
        <div className="max-w-2xl border mx-auto text-3xl mt-60
text-center">
            Data-Art
        </div>
    );
}
```

The css styling is declared within a div attribute called className. First, we set the maximum width and the border, then create a margin along the *x* axis (margin left and right to be auto), declare the font size to be text-3xl, set the margin top to be 60, and then center the text within the div instance.

Based on the styling, we should see the following output:

Data-Art

Figure 8.4 – Centering the div and text

The code base is set and we are ready to implement our no-code environment.

In this section, we saw how to set up a React environment for our app. We also saw how to configure `tailwindcss` for our app. In the next section, we will learn how to structure and design the app.

# Structuring and designing the app

React.js has some core philosophy of app design, which is mostly about breaking up the UI into a component hierarchy, and also one of the ideas is to identify where your state should live.

In this section, we will see how to design the structure of our no-code app with React.js and also consider the React philosophy of app design. With this principle, we will find it easy to implement a basic UI in React.

First, let's understand what a no-code environment is and what we want to achieve with it. The no-code environment is used to make data handling and analysis easier with just the click of a few buttons.

We will create a platform where users can upload their data, perform analysis, and do what they do with code, such as the following:

- DataFrame-to-DataFrame operations such as `concat`
- Arithmetic operations such as `cummax` and `cumsum`
- Querying to filter out a DataFrame by a column value
- Describing a DataFrame

We want to be able to do all that without actually coding, and everything will be done in the browser. We also want to make it possible to get insights from data via data visualization using a bar chart, line chart, and pie chart. The following figure shows a sketch of the app design:

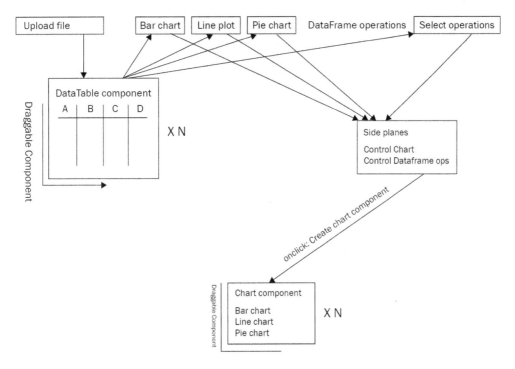

Figure 8.5 – App structure and design sketch

*Figure 8.5* shows the structure and design of the app. The app is divided into three main components, as follows:

- The `Navbar` component, containing file upload, a bar chart, a line chart, and the `DataFrame` operation select field
- The main body containing a `Data Table` component and a `chart` component
- `SideBar`, containing the side planes for chart and `DataFrame` operations

The app workflow can be described as follows:

1. First, a data file (`csv`) is uploaded.
2. By uploading a file, the first `Data Table` is created. This is a component containing the display of DataFrame.
3. To perform any operation, such as `DataFrame` operations or chart operations, the Data Table is selected, so that we can identify the correct table to perform operations on.

4.  For chart operations, either a bar chart, line chart, or pie chart is clicked. This clicked event activates the `Side Plane` for chart operations.

5.  If a `DataFrame` operation is selected, the `Side Plane` is activated for `DataFrame` operations.

6.  When you've filled in the necessary fields in the `Side Plane`, a new chart component and `Data Table` component are created.

The following figure describes the whole workflow:

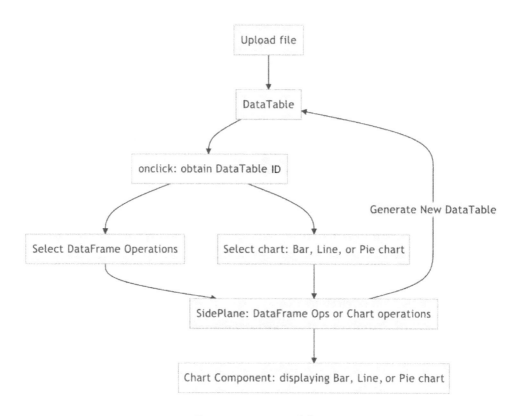

Figure 8.6 – App workflow

The workflow shows how each component responds to one another. For instance, without uploading a file, the main body and the `Side Plane` will be invisible. Even with file upload, the `Side Plane` remains hidden and only comes up whenever a DataFrame or chart operation is to be performed on a particular data table.

With this, it shows that we need to create a state to manage the activation of the main body whenever a file is uploaded and also a state to manage how the `Side Plane` is activated when operations are being done on a data table. Also, note that the `Side Plane` contains two operations, and we must display the fields for these operations based on the type of operations being selected.

If a chart operation is being selected, the `Side Plane` needs to display the necessary fields for the selected plot chart, either bar, line, or pie chart, and if it's a `DataFrame` operation that's being selected, the `Side Plane` needs to display the DataFrame operation fields as shown in the following figure:

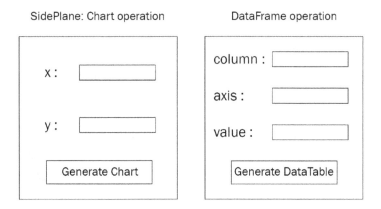

Figure 8.7 – Side plane operation fields

From *Figure 8.7*, we can see that the data table and `chart` components have an **x** and **y** symbol. This shows that we can have more than one data table or chart component and each of these components is draggable. Hence, we need to create a state to manage the list of `Data Table` components and `chart` components. Having a state for each component makes it possible to create, update, and delete the component as we desire.

As described in the app workflow, the `Side Plane` operations need data for visualization and `DataFrame` operations and this data is obtained by clicking on our desired `Data Table` to work on. Each `Data Table` stores its own DataFrame object (we'll look into this deeply while implementing the steps). Hence, whenever a data table is clicked, its index in the data table state is obtained and passed into the side plane alongside the data table state that indicates which data table to work on.

Also, for the side plane to know which chart type (bar, line, or pie) is required for the operation or what type of `DataFrame` operation is to be done, we create a state to manage what type of chart or `DataFrame` is currently selected.

In summary, the set of states that's needed is described as follows:

- State to manage the `DataTable` list
- State to manage the list of charts
- State to show to manage the visibility of `SidePlane`
- State to manage the current `DataTable` index
- State to manage the type of chart selected
- State to manage the current `DataFrame` operation selected

The state created here is not well optimized. It's possible to manage the number of states created; for example, the same state managing `Side Plane` visibility can also be used to manage the type of chart selected.

Since we will be using more than one or two states and some of the states interact with another state, we could use `useReducer` (a React Hook) to manage the state interactions but we would like to make this simple without adding overhead knowledge.

In this section, we talked about the app design and structure. We also designed the app workflow and talked about different states to be created for the app. In the next section, we will talk about the app layout and `DataTable` components. We will see how to create a data table component and how to manage the state. We will also look into uploading a file in the browser with Danfo.js.

# App layout and the DataTable component

In this section, we'll see how to lay out the app based on the design and workflow discussed in the previous section. Also, we will implement the `DataTable` component, responsible for the display of `DataFrame` tables. We will also implement the `DataTables` component, responsible for displaying different `DataTable` components.

We've seen the sketch of what the app will look like and also seen the basic workflow of the app. We will start to implement the steps by first building the basic layout of the app and then implementing the data table component.

With `tailwindcss`, it's quite easy to lay out the app. Let's create a file called `App.js` and input the following code:

```
function App() {
  return (
    <div className="max-w-full mx-auto border-2 mt-10">
      <div className="flex flex-col">
        <div className="border-2 mb-10 flex flex-row">
          Nav
        </div>
        <div className="flex flex-row justify-between
border-2">
          <div className="border-2 w-full">
            <div>
              Main Body
            </div>

          </div>
          <div className="border-2 w--1/3">
            Side Plane
          </div>
        </div>
      </div>

    </div>
  );
}
```

The preceding code snippet creates the layout of the app with a `flex` box. This layout shows the basic components of the app, which are `Nav`, `Main Body`, and `Side Plane`.

> **Note**
>
> The `css` instance being used in the tutorial won't be explained. Our main focus is on building the functionality of the app.

If everything runs well, we should get the following output:

Nav

Main Body                                                                    Side Plane

Figure 8.8 – App layout

We've laid out the app. Let's move on to implementing `DataTable` components to display the result of all `DataFrame` operations.

## Implementing DataTable components

A `DataTable` component is responsible for the display of a data table. For each DataFrame operation, we generate a new data table showing the result of the operation as shown in the following sketch:

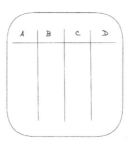

Figure 8.9 – Data table

For the display of the table, we'll make use of a React package called `react-table-v6` and since we want the `DataTable` component to be draggable across the page, there is a package called `react-draggable`, which makes it easier to implement the feature.

> **Note**
> The code for `DataTable` can be pulled from here: `https://github.com/PacktPublishing/Building-Data-Driven-Applications-with-Danfo.js/blob/main/Chapter08/src/components/DataTable.js`.

We need to add these packages to our code base using `yarn`:

```
yarn add react-table-v6 react-draggable
```

Once the packages are installed, let's create a `DataTable` component in the `src/component/DataTable.js` file with the following steps:

1. We import the necessary packages:

```
import React from "react";
import ReactTable from 'react-table-v6'
import Draggable from 'react-draggable';
import 'react-table-v6/react-table.css'
```

2. We create the `DataTable` component:

```
export default function DataTable({ columns, values,
setCompIndex, index }) {
    // DataTable component code here
}
```

The `DataTable` component takes in the following `props` values:

- `columns`: Data table column names.
- `values`: Data table values for each column.
- `setCompIndex`: This is a state function used to manage the current data table selected.
- `index`: This is the index of the current table.

3. For the `react-table` component, we need to reshape the column and the values to fit the desired input for the `react-table` component.

Let's reshape the column value to be passed into `react-table`:

```
const dataColumns = columns.map((val, index) => {
    return { Header: val,
      accessor: val,
      Cell: (props) => (
        <div className={val || ''}>
          <span>{props.value}</span>
        </div>
      ),
      width:
        index === 0 && (1280 * 0.8333 - 30) / columns.
length < 130
        ? 130
```

```
        : undefined,
    }
});
```

Using the preceding code, the column names, which are the `Header` for the table, are transformed to the following shape:

```
[{
  Header: "A",
  accessor: "A"
},{
  Header: "B",
  accessor: "B"
},{
  Header: "C",
  accessor: "C"
}]
```

The `Header` key is the name of the column to be shown in the table, and `accessor` is the key in the data.

4.  We need to transform the data table values into a format needed by `react-table`. The following code is used to transform data table values:

```
const data = values.map(val =>{
    let rows_data = {}
    val.forEach((val2, index) => {
        let col = columns[index];
        rows_data[col] = val2;
    })
    return rows_data;
})
```

As shown in the preceding code, we will transform the data table values into the following data form:

```
[{
  A: 2,
  B: 3,
  C: 5
},{
```

```
    A: 1,
    B: 20,
    C: 50
}, {
    A: 23,
    B: 43,
    C: 55
}]
```

Initially, `values` is an array of an array that is being transformed into the preceding data format and then assigned to the `data` variable.

The accessor declared in the preceding column format points to the value of each of the keys in the dictionary. Sometimes, we might have nested data in the following format:

```
[{
    dummy: {
        A: 1.0,
        B: 3.0
    },
    dummy2: {
        J: "big",
        k: "small"
    }
}, . . . . ]
```

For this type of data format, we can declare the `data` column to be in the following format:

```
[{
    Header: "A",
    accessor: "dummy.A"
},
{
    Header: "B",
    accessor: "dummy.B"
},
{
    Header: "J",
```

```
    accessor: "dummy2.J"
  },
  {
    Header: "K",
    accessor: "dummy2.K"
  }]
```

For this project, we won't be working with this nested data format, so there won't be a need to dive deeper, but if you are curious, you can check the `react-table-v6` documentation.

The column names, including `Header`, and the table data are now in the right format and are ready to be passed into the `react` table. The `DataTable` component is now updated to contain the following code:

```
function DataTable({ columns, values, setCompIndex, index }) {
  . . . . . . . . . . . . . . . . . . .
  const handleSidePlane = ()=>{
      setCompIndex(index)
  }
    return (
      <Draggable >
          <div className="w-1/2" onClick={ ()=>
handleSidePlane() }>
            <ReactTable
              data={data}
              columns={dataColumns}
              getTheadThProps={() => {
                return { style: { wordWrap: 'break-word',
whiteSpace: 'initial' } }
              }}
              showPageJump={true}
              showPagination={true}
              defaultPageSize={10}
              showPageSizeOptions={true}
              minRows={10}
            />
          </div>
```

```
        </Draggable>
    )
}
```

The `ReactTable` component is wrapped in the `Draggable` component to make the `DataTable` component draggable. In the `ReactTable` component, we set up some pagination fields, such as setting the default page to be `10`.

Recall how when designing the workflow of the app we mentioned tracking the ID of a `Data Table` when clicked. The `handleSide Plane` function is used to make a call to `setCompIndex`. `setCompIndex` is used to update the `compIndex` state, which stores the index of the selected `Data Table`.

> **Note**
>
> The code for `DataTables` is available here: `https://github.com/PacktPublishing/Building-Data-Driven-Applications-with-Danfo.js/blob/main/Chapter08/src/components/DataTables.js`.

Several Data Tables will be generated per operation, hence we need to manage the display of this `Data Table`. We'll create a component that manages the display of all the `Data Tables` generated; hence, we'll create a file in the component directory and name it `Data Tables`, containing the following code:

```
import React from 'react'
import DataTable from './DataTable'
export default function DataTables({datacomp, setCompIndex,}) {
    return (
        <div>
            {datacomp.map((val,index) => {
                return(
                        <>
                        <DataTable
                            key={index}
                            columns={val.columns}
                            values={val.values}
                            setCompIndex={setCompIndex}
                            index={index}                      />
                        </>
```

```
      )
    })}
  </div>
  )
}
```

This component loops through the `datacomp` state and passes each prop into the `DataTable` components.

In the next sub-section, we'll go ahead and initialize different states and also show how to upload a CSV and obtain our data.

# File upload and state management

From the app design, we see that for any operation to take place, we need to upload a file first. By uploading the files, we create the `DataFrame` that will be used by the `DataTable` and `chart` components.

> **Note**
>
> Across this chapter, the code in `App.js` is gradually updated based on the implementation of new components. But the final code for `App.js` is available here: `https://github.com/PacktPublishing/Building-Data-Driven-Applications-with-Danfo.js/blob/main/Chapter08/src/App.js`.

We'll update the code in `App.js` with the following step to contain the `Data` component state, file upload, and updating of the state:

1. We import `React` and a React hook called `useState`:

   ```
   import React, { useState } from 'react';
   ```

2. We import the `read_csv` method, which will be used to read the CSV files uploaded:

   ```
   import { read_csv } from 'danfojs/src/io/reader' // step
   2
   ```

3. We create a state to store a list of data for each `DataTable` component generated:

   ```
   const [dataComp, setDataComp] = useState([])
   ```

4.  We then create a function to manage file upload and read the uploaded file into a
    `DataFrame`:

```
const changeHandler = function (event) {
    const content = event.target.files[0]
    const url = URL.createObjectURL(content)

    read_csv(url).then(df => {
      const columns = df.columns
      const values = df.values
      setDataComp(prev => {
        let new_data = prev.slice()
        let key = new_data.length + 1
        let dict = {
          columns: columns,
          values: values,
          df: df,
          keys: "df" + key
        }
        new_data.push(dict)
        return new_data
      })

    }).catch((error) => {
      console.log(error)
    })
}
```

In the preceding code, we generate a blob URL from the uploaded file with `URL.createObjectURL`. This is done because the `read_csv` code in `Danfo.js` is only taken in the local path of the CSV file, the HTTP URL of a CSV file, and the blob URL of a CSV file.

The generated URL is then passed into the `read_csv` function. Since `read_csv` is an asynchronous function, we need to wait for the promise to be resolved, and then collect the returned value from the promise via the `then` method. The return value from the resolved promise is a `DataFrame`.

With `read_csv`, the CSV data is converted into a `DataFrame` and then the `DataComponent` state is updated. Using the `setDataComp` state function, we created an object containing the following keys:

a) `columns`: To store the headers (column name) of the CSV file

b) `values`: To store the CSV data point, which is the `DataFrame` values

c) `df`: To store the `DataFrame` generated

d) `keys`: Generates a key for each of the Data component `data`

There was a decision to be made, between actually saving the DataFrame itself in the state data for each component. Since we have the column name and the `DataFrame` value stored, it looks redundant.

But the reason we finally go with storing it is that it will be computationally expensive to always create a `DataFrame` from the column and values every time we need to perform `DataFrame` operations.

Also, `columns` and `values` are stored for easy accessibility whenever we want to generate a table from the `react-table` component. But still, it feels redundant and as a personal exercise (among the list of to-dos that will be listed at the end of this section), you can go ahead and clean it up.

5.   We print out the output of the `dataComp` state in the browser console once the state is updated:

```
if (dataComp.length) { //step 8
    console.log("dataComp column", dataComp[0].columns)
    console.log("dataComp values", dataComp[0].values)
    console.log("dataComp dataFame", dataComp[0].df)
}
```

The following screenshot shows the updated UI of the app:

Choose File  No file chosen

Main Body                                                    Side Plane

Figure 8.10 – Updated UI for file upload

Once a file is uploaded, we should see the following output in the browser console:

```
dataComp column ▶ (3) ["A", "B", "C"]
dataComp values ▼ (5) [Array(3), Array(3), Array(3), Array(3), Array(3)] ⓘ
                  ▶ 0: (3) [1, 2, 20]
                  ▶ 1: (3) [2, 3, 10]
                  ▶ 2: (3) [3, 3.2, 30]
                  ▶ 3: (3) [4, 5, 60]
                  ▶ 4: (3) [5, 2, 70]
                    length: 5
                  ▶ __proto__: Array(0)
dataComp dataFame ▶ DataFrame {kwargs: {…}, series: false, data: Array(5), row_da
```

Figure 8.11 – dataComp state output

We've set up the file upload and also the state management for each of the DataTable components. Let's integrate the DataTables component created into the app.

## Integrating the DataTables component into App.js

App.js will be updated with the following steps:

1.  We import the DataTables component and create a compIndex state, which enables us to store the index of the DataTable component we want to work on at the moment:

```
. . . . . . . . . . .
import DataTables from './components/DataTables';
function App() {
    . . . . . . . . . . .
    const [compIndex, setCompIndex] = useState()
    . . . . . . . . . .
}
```

2.  We then add the DataTables component to the App component:

```
<div>
    { (dataComp.length > 0) &&
        <DataTables
            datacomp={dataComp}
            setCompIndex={setCompIndex}
        />
    }
</div>
```

To enable `DataTable` component visibility, we check that the `dataComp` state is not empty. Before uploading a file, if the `dataComp` state is empty, the `DataTable` component will not be visible. Once a file is updated, the `DataTable` component becomes visible, since the `dataComp` state is no longer empty.

The preceding code should give us the following output once a file is uploaded:

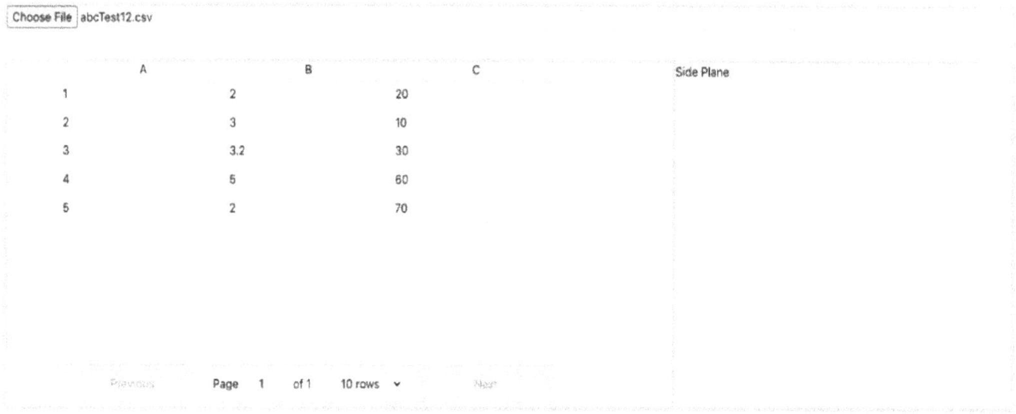

Figure 8.12 – Display of the DataTable component on file upload

In this section, we talked about file upload and `DataTable` creation and management and saw how to manage state. In the next section, we will implement different `DataFrame` operation components and also implement the `Side Plane` for `DataFrame` operations.

# Creating different DataFrame operation components

In this section, we will create different `DataFrame` operation components and also implement the `Side Plane` for `DataFrame` operation components. Danfo.js contains a lot of `DataFrame` operations. If we were to design a component for each, it would be very stressful and redundant.

To prevent the creation of a component for each `DataFrame` method, we group each of the `DataFrame` operations based on their (keyword) argument, that is, based on the variable passed into them. For example, there are some `DataFrame` methods that take in only the axis of operation, hence we can group these types of methods together.

Here is a list of `DataFrame` operation components to be created and the `DataFrame` method grouped under them:

- **The Arithmetic component**: This contains the `DataFrame` method whose argument is only the axis of operation, which can be either `1` or `0`. The methods used to carry out arithmetic operations on `DataFrame` include `min`, `max`, `sum`, `std`, `var`, `sum`, `cumsum`, `cummax`, and `cummin`.

- **The Df2df component**: This contains the operation between two `DataFrame` components, such as the logical operation between a `DataFrame` and a series, a value, or a `DataFrame`. The methods used to carry out these operations are `concat`, `lt`, `gte`, `lte`, `gt`, and `neq`.

- **The Query component**: This is a component for a `DataFrame` query.

- **The Describe component**: This is a component for describing the `DataFrame` statistic.

We will start looking at the implementation of these components starting from the least complex in this order: `Describe`, `query`, `Df2df`, and `Arithmetic`.

# Implementing the Describe component

In this section, we will implement the `Describe` component and also integrate the `Side Plane` component.

In the `src/Components/` directory, let's create another folder called `Side Planes`. This folder will contain all the components for `DataFrame` operations.

In the `Side Planes/` folder, let's create a `".js"` file named `Describe.js` and update it with the following steps:

1. We create the `Describe` functional component, taking in the `dataComp` state and the `setDataComp` state function, to update the `dataComp` state with the generated `DataFrame`:

   ```
   export default function Describe({ dataComp,
   setDataComp}) {

   }
   ```

2. We create a button named `Describe`:

   ```
   return (
       <div>
           <button onClick={ ()=> describe()} className="bg-
   ```

```
blue-700 text-white rounded-sm p-2">Describe</button>
    </div>
)
```

The Describe component has a button interface because it does not take in any arguments. The button has an onClick event that triggers the Describe function whenever the button is clicked.

3. We then implement the describe() function, which is triggered anytime the describe button is clicked:

```
const describe = ()=> {
    const df = dataComp.df.describe()
    let column = df.columns.slice()
    column.splice(0,0, "index")
    const values = df.values
    const indexes = df.index

    const new_values = values.map((val, index)=> {
        let new_val = val.slice()
        new_val.splice(0,0, indexes[index])
        return new_val
    })
    . . . . . . . . .
}
```

We obtain the df key containing the DataFrame from the dataComp state and then call the describe method.

From the DataFrame generated from the describe operation, we obtain the columns and add an index to the list of column names. The index is added at the start of the list; this is done because the index generated from the describe method is needed to capture what each row in the data is.

Next, we obtain the DataFrame value, loop through it, and add the index values to the DataFrame value obtained.

4. We update the dataComp state with the new generated column, value, and DataFrame:

```
setDataComp(prev => { // step 7
    let new_data = prev.slice()
```

```
        let dict = {
            columns: column,
            values: new_values,
            df: df
        }
        new_data.push(dict)
        return new_data
    })
```

To view this component in action, we will need to implement the DataFrame operation selection field, as shown in the App design sketch in *Figure 8.5*. This DataFrame operation selection field enables us to select which of the DataFrame operation components to show in Side Plane.

To do this, we need to add the select field for the DataFrame operation in the Navbar component alongside the input field for file upload. Also, we need to implement condition rendering for each of the DataFrame operation components shown in Side Plane.

## Setting up SidePlane for the Describe component

In the Side Planes/ folder, let's create a file called Side Plane.js and input the following code:

```
import React from 'react'
import Describe from './Describe'
export default function SidePlanes({dataComp,
  dataComps,
  setDataComp,
  df_index,
  dfOpsType}) {

    if(dfOpsType === "Arithemtic") {
      return <div> Arithmetic </div>
    }
    else if(dfOpsType === "Describe") {
      return <Describe
          dataComp={dataComp}
          setDataComp={setDataComp}
```

```
        />
    }
    else if(dfOpsType === "Df2df") {
      return <div> Df2df </div>
    }
    else if(dfOpsType === "Query") {
      return <div> Query </div>
    }
  return (
    <div>
      No Plane
    </div>
  )
}
```

In the preceding code, we create a Side Plane component. This component contains conditional rendering based on the type of data operation selected. The DataFrame operation selected is managed by the dfOpsType state.

The Side Plane takes in the dataComp state, which can be any of the data stored in the dataComps state. Some DataFrame operations will need the selected dataComp state and also the whole state, which is dataComps state, for their operation.

In the Side Plane component, we will check dfOpsType to find out the type of operation passed and the interface to render in the side plane.

Before we move into integrating Side Plane into App.js, let's create an index.js file in the Side Planes/ folder. With this, we can define the component to import. Since we are using conditional rendering from the Side Plane component, we just need to export the Side Plane component in the index.js instance, as shown in the following code:

```
import SidePlane from './SidePlane'
export { SidePlane }
```

The preceding code enables us to import Side Plane in App.js.

## Integrating SidePlane into App.js

Side Plane is created. Let's integrate it into App.js and also add the HTML select field for the DataFrame operation to App.js, as shown in the following code:

1. We import the SidePlane component:

```
import { SidePlane } from './components/SidePlanes'
```

2. We update the App component functionality with the following code:

```
function App () {
    . . . . . . . .
  const [dfOpsType, setDfOpsType] = useState() // step 2
  const [showSidePlane, setSidePlane] = useState(false)
//step 3
    . . . . . . . .
  const dataFrameOps = ["Arithemtic", "Describe",
"Df2df", "Query"] // step 4
  const handleDfops = (e) => { //step 6
    const value = e.target.value
    setDfOpsType(value)
    setSidePlane("datatable")
  }
    . . . . . . . . . . . .
  }
```

In the preceding code, we create the dfOpsType state to store the current type of DataFrame operation selected.

showSidePlane is also created to manage SidePlane visibility. Also, an array of DataFrame operations is created. We then create a function to handle updating dfOpsType and the showSidePlane state whenever a DataFrame operation is clicked.

3.  We then add the `SidePlane` component:

```
<div className="border-2 w-1/3">
    {showSidePlane
        &&
        (
        showSidePlane === "datatable" ?
            <div className="border-2 w-1/3">
                <SidePlane
                    dataComp={dataComp[compIndex]}
                    dataComps={dataComp}
                    df_index={compIndex}
                    setDataComp={setDataComp}
                    dfOpsType={dfOpsType}
                />
            </div> :
            <div className="border-2 w-1/3">
                Chart Plane
            </div>
        )
    }
</div>
```

In the preceding code, we display the `SidePlane` by first checking that the `SidePlane` state is not *false*, then we check the type of `SidePlane` to display. Since we've only implemented the `Describe` component among the list of `DataFrame` operations, let's upload a file and then perform a `DataFrame` operation. The following screenshot shows the result of performing a `Describe` operation on a `DataTable`:

Figure 8.13 – Describe operation on DataTable

In the preceding screenshot, the data table at the top left is generated when the file is uploaded and the DataFrame at the bottom right as a result of the Describe operation.

In this section, we saw how to implement the Describe component and how to manage Side Plane visibility. In the next section, we will implement the Query component for the Query method in the DataFrame.

## Implementing the Query component

In this section, we'll create a component for the DataFrame query method. This component will aid the filtering of the DataFrame by column values per Data Table.

> **Note**
>
> The code for a Query component is available here: https://github.
> com/PacktPublishing/Building-Data-Driven-
> Applications-with-Danfo.js/blob/main/Chapter08/
> src/components/SidePlanes/Query.js.

Let's create a file in the `components/Side Planes/` folder named `Query.js` and update it with the following steps:

1. We create the `Query` component:

```
import React, { useRef } from 'react'

export default function Query({ dataComp, setDataComp}) {
  // step 1
  const columnRef = useRef()
  const logicRef = useRef()
  const valuesRef = useRef()
  . . . . . . . . . . . .
}
```

We created a `useRef` hook variable, which enables us to obtain the current value of the input to the following input fields: the column field (takes in the name of the column to query), the logic field (takes in the logic value to be used for the query), and the value field (takes in the value to be used to query the selected column).

2. We then update the `Query` component with the following code:

```
const columns = dataComp.columns
const logics = [">", "<", "<=", ">=", "==", "!="]
```

In the preceding code, we obtain the column names available in the `DataFrame` of the current `Data Table`. This column name will be used to populate the select field to which the user can select the column to query.

We also create a list of symbols characterizing the type of logic operation we want to perform. This symbol will also be used to populate a select field to which users can select the logical operation to be used for the query.

3. Create a `query` function. This function will be triggered to perform the query operation whenever the **query** button is clicked:

```
const query = ()=>{
    const qColumn = columnRef.current.value
    const qLogic = logicRef.current.value
    const qValue = valuesRef.current.value

    const  df = dataComp.df.query({column: qColumn, is:
qLogic, to: qValue})
```

```
        setDataComp(prev => {
          let new_data = prev.slice()
          let dict = {
            columns: df.columns,
            values: df.values,
            df: df
          }
          new_data.push(dict)
          return new_data
        })
      }
```

We obtain the value of each of the input fields (select fields) whenever the `query` function is triggered. For example, to obtain the value of the column field, we make use of `columnRef.current.value`. The same thing is applicable for obtaining the value in another field.

We also invoke the query method for the `DataFrame` belonging to the current `dataComp` state. The value obtained from each of the input fields is passed into the query method to perform the operation.

The `dataComps` state is updated by using the `setDataComp` state function. By updating the `dataComps` state, a new `DataComp` state is created containing the result of the `query` method.

## Implementing the query component interface

We've seen the backend of the `Query` component, so now let's build an interface for it. Let's update the preceding code in `Query.js` with the following steps:

1. For the query UI, we create a form containing three different input fields. First, we create the input field for the column field:

```
<div>
  <span className="mr-2">Column</span>
  <select ref={columnRef} className="border">
    {
      columns.map((column, index) => {
        return <option value={column}>{column}</option>
      })
    }
```

```
      </select>
  </div>
```

For the column field, we loop through the column array to create HTML select field options for the list of columns in the `DataFrame`. We also include `columnRef` to track the selected column name.

2. We then create the logic input field:

```
<div>
  <span className="mr-2">is</span>
  <select ref={logicRef} className="border">
    {
      logics.map((logic, index) => {
        return <option value={logic}>{logic}</option>
      })
    }
  </select>
</div>
```

We loop through the `logic` array and fill in the HTML select field with the logic operators. Also, `logicRef` is added to the HTML select field to obtain the selected logic operator.

3. We then create the `input` field for the query value:

```
<div>
  <span className="mr-2">to</span>
    <input ref={valuesRef} placeholder="value"
className="border"/>
</div>
```

4. We create a `button` class name to make a call to the `query` function:

```
<button onClick={()=>query()} className="btn btn-default
dq-btn-add">Query</button>
```

To visualize the `query` component in the main app, let's update the `SidePlane` component in `SidePlane.js`:

```
Previous code:
. . . . . . .
else if(dfOpsType === "Query") {
```

```
        return <div> Query </div>
    }
Updated code:
else if (dfOpsType === "Query") {
        return <Query
                dataComp={dataComp}
                setDataComp={setDataComp}
        />
    }
```

The preceding code updates the `Side Plane` to contain the `Query` component. If we perform a query operation on an uploaded file, we should get the following result:

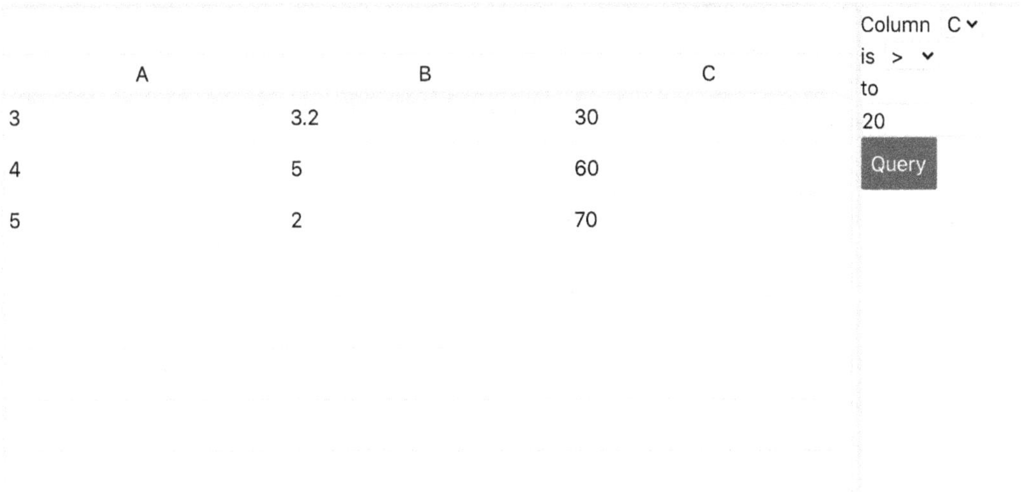

| | A | B | C | Column C ⌄ |
|---|---|---|---|---|
| | | | | is > ⌄ |
| | | | | to |
| 3 | 3.2 | 30 | | 20 |
| 4 | 5 | 60 | | Query |
| 5 | 2 | 70 | | |

Figure 8.14 – Query operation run on column C, checking whether its value is greater than 20

In this section, we created a `query` component. In the next section, we will look into creating a component for operations involving `DataFrame-to-DataFrame` operations, series, and scalar values.

# Implementing the Df2df component

In this section, we will be implementing a component for performing an operation between a `DataFrame` and another `DataFrame`, `Series`, and `Scalar` values.

> **Note**
>
> The code for the `Df2df` component is available here: `https://github.com/PacktPublishing/Building-Data-Driven-Applications-with-Danfo.js/blob/main/Chapter08/src/components/SidePlanes/Df2df.js`.

There are different components in Danfo.js that perform the operation between a `DataFrame` and a `Series` and a `DataFrame` and a `Scalar` value. To prevent having to create a component for each of these methods, we can group them together to form a single component.

A list of `DataFrame` methods we plan on grouping together is as follows:

- Less than (`df.lt`)
- Greater than (`df.gt`)
- Not equal to (`df.ne`)
- Equal to (`df.eq`)
- Greater than or equal to (`df.ge`)
- Addition (`df.add`)
- Subtraction (`df.sub`)
- Multiplication (`df.mul`)
- Division (`df.div`)
- Power (`df.pow`)

One common attribute in the preceding list of methods is that they all take in the same type of arguments, which is the value (it can be a `DataFrame`, `Series`, or `scalar` value) and the axis on which the operation is to be performed.

If we look at the `DataFrame` concat method, it also takes in arguments of the same pattern similar to the methods in the preceding list. The only difference is that for the `concat` method, the `df_list` argument is an array of `DataFrames`.

Let's create a file in the `Side Planes/` folder named `Df2df.js`. In this file, we will implement the `Df2df` component with the following steps:

1.  First, we import `concat` from Danfo.js, and then create the `Df2df` component:

```
import React, { useRef } from 'react'
import { concat } from 'danfojs/src/core/concat'

export default function Df2df({dataComp, dataComps,df_
index, setDataComp}) {
  const dfRef = useRef()
  const inpRef = useRef()
  const axisRef = useRef()
  const opsRef = useRef()

  const allOps = [
    "lt", "ge", "ne",
    "eq", "gt", "add",
    "sub", "mul", "div",
    "pow", "concat"
  ]
  . . . . . . . . . .  . . . .
}
```

We created a reference variable for each of the input fields. For the `Df2df` operation, we have four input fields (the `DataFrame` selection field, the `scalar` value input field, the `axis` field, and the `operation` type field).

The `operation` type fields contain the list of all operations available in the `Df2df` component. This will be a select field, hence users can select any of the operations to work with.

We also create an `allOps` list of all operations to be offered by the `Df2df` component.

2.  We also need to create a function to perform the `Df2df` operation whenever the `submit` button is clicked:

```
const df2df = () => {
  // step 4
  let dfIndex = dfRef.current.value
```

```
let inp = parseInt(inpRef.current.value)
let axis = parseInt(axisRef.current.value)
let ops = opsRef.current.value
. . . . . . . . . . . . .
}
```

We obtained values from all the reference variables belonging to each input field.

3. We update the df2df function with the following code:

```
if( ops != "concat") {
    let value = dfIndex === "None" ? inp :
dataComps[dfIndex].df
    let df = dataComp.df

    let rslt = eval('df.${ops}(value, axis=${axis})')
// step 6

    setDataComp(prev => {
      let new_data = prev.slice()
      let key = new_data.length +1
      let dict = {
        columns: rslt.columns,
        values: rslt.values,
        df: rslt,
        keys: "df" + key
      }
      new_data.push(dict)
      return new_data
    })
  }
```

We check that the operation selected is not a concat operation. This is done because concat operations take in a list of DataFrames instead of just a DataFrame or Series.

We make use of the eval function to prevent writing multiple if conditions to check which DataFrame operation to call.

4.  We implement the condition for the `concat` operation. We also make a call to the `concat` method in the `DataFrame`:

```
. . . . . . . . .
else { // step 7
        let df2 = dataComps[dfIndex].df
        let df1 = dataComp.df
        let rslt = concat({ df_list: [df1, df2], axis: axis
})

        let column = rslt.columns.slice()
        column.splice(0,0,"index")
        let rsltValues = rslt.values.map((val, index) => {
          let newVal = val.slice()
          newVal.splice(0,0, rslt.index[index])
          return newVal
        })
        . . . . . . . . . . .
}
```

The preceding steps show the backend implementation of the `Df2df` component.

## Implementing the Df2df component interface

Let's update the code for the UI with the following steps:

1.  For the UI, we need to create a form containing four input fields. First, we create an input field to select the type of `DataFrame` operation we want:

```
<div>
    <span className="mr-2"> Operations</span>
    <select ref={opsRef}>
      {
          allOps.map((val,index) => {
          return <option value={val} key={index}>{val}</
    option>
          })
      }
    </select>
</div>
```

We loop through the `allops` array to create an `input` field to select different types of `DataFrame` operations.

2.  We then create an `input` field to select the `DataFrame` we want to perform the operation selected on:

```
<div>
  <span className="mr-2"> DataFrames</span>
  <select ref={dfRef}>
    <option key={-1}>None</option>
    {
      dataComps.map((val,index) => {
        if( df_index != index) {
          return <option value={index}
key={index}>{'df${index}'}</option>
        }
      })
    }
  </select>
</div>
```

We also loop through the `dataComps` state to obtain all `dataComp` state in it except for the `dataComp` state we are performing the operation with.

3.  We then create an `input` field to input our value; in this case, we are performing an operation between `DataFrame` and a `Scalar` value:

```
<div>
  <span>input a value</span>
  <input ref={inpRef} className="border" />
</div>
```

4.  We create an `input` field to select the axis of operation:

```
<div>
  <span>axis</span>
  <select ref={axisRef} className="border">
    {
      [0,1].map((val, index) => {
        return <option value={val} key={index}>{val}</
option>
```

```
        })
      }
    </select>
  </div>
```

5.  We then create a button that triggers the df2df function to perform a Df2df operation based on the input fields:

```
<button onClick={()=>df2df()} className="bg-blue-500 p-2
text-white rounded-sm">generate Dataframe</button>
```

In the preceding steps, we created the UI for the component.

Let's update the SidePlane component to contain the Df2df component:

```
import Df2df from './Df2df'
export default function SidePlanes({dataComp,
  dataComps,
  setDataComp,
  df_index,
  dfOpsType}) {
    . . . . . . .
    else if(dfOpsType === "Df2df") {
      return <Df2df
          dataComp={dataComp}
          dataComps={dataComps}
          df_index={df_index}
          setDataComp={setDataComp}
      />
    }
    . . . . . . . .


    }
```

The preceding code adds the Df2df component to the SidePlane component and passes the required props in the Df2df component. The following screenshot shows the upload of two CSV files with the same content:

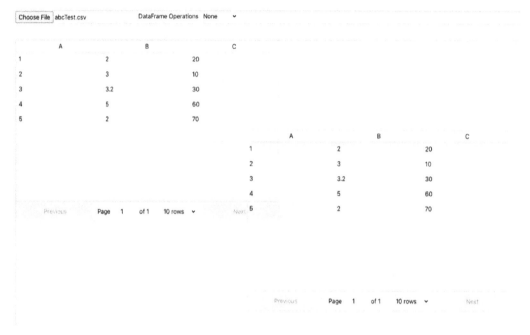

Figure 8.15 – Uploading CSV files with the same content

The following shows the output of performing a Df2df operation (the concat operation to be specific) on the selected Data Table:

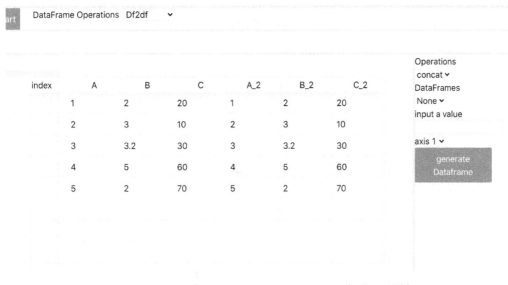

Figure 8.16 – Performing a concat operation on the Data Table

In this section, we created the Df2df component for performing an operation between two DataFrames, and between a DataFrame and a Series/Scalar value.

In the next section, we will implement the last DataFrame component, which is the arithmetic component for the DataFrame arithmetic operation.

# Implementing the Arithmetic component

We will implement the arithmetic component to perform some of the arithmetic operations provided in Danfo.js.

---

**Note**

The code for the Arithmetic component is available here: https://github.com/PacktPublishing/Building-Data-Driven-Applications-with-Danfo.js/blob/main/Chapter08/src/components/SidePlanes/Arithemtic.js.

---

Let's create a file named `Arithmetic.js` in the `Side Planes/` folder. The following steps will be used to create the `Arithmetic` component:

1. We create an `Arithmetic` component:

```
import React, { useRef } from 'react'
export default function Arithmetic({ dataComp,
setDataComp}) {

  const seriesOps = ["median", "min", "max", "std",
"var", "count", "sum"]
  const dfOps = ["cumsum", "cummax", "cumprod", "cummin"]
  const all = ["median", "min", "max", "std", "var",
"count", "sum",
              "cumsum", "cummax", "cumprod", "cummin"]

  const axisRef = useRef()
  const opsRef = useRef()
  . . . . . . . . . . .
}
```

We create different arrays to store different operations, such as `seriesOps` for a Series operation and `dfOps` for a DataFrame operation. We also create an `all` array that stores all these operations (`Series` and `DataFrame`) together.

2. We create a function called `arithmetic`. This function is used to perform the arithmetic operations:

```
const arithemtic = () => {

    let sOps = opsRef.current.value
    let axis = axisRef.current.value
    if ( seriesOps.includes(sOps)) {
      let df_comp = dataComp.df
      let df = eval('df_comp.${sOps}(axis=${axis})')

      let columns = Array.isArray(df.columns) ?
df.columns.slice() : [df.columns]
      columns.splice(0,0, "index")
```

```
let values = df.values.map((val,index) => {

    return [df.index[index], val]
})
. . . . . . . . . . .
}
```

We obtain the values from the input fields, `opsRef.current.value` and `axisRef.current.value`. We also check whether the operation selected belongs to `seriesOps`. If so, we perform the operation selected.

3.  We perform a `DataFrame` operation if the operation does not belong to `seriesOps`:

```
else {

    let df_comp2 = dataComp.df
    let df = eval('df_comp2.${sOps}({axis:${axis}})')

    setDataComp(prev => {
      let new_data = prev.slice()
      let dict = {
        columns: df.columns,
        values: df.values,
        df: df
      }
      new_data.push(dict)
      return new_data
    })
}
```

The preceding steps are used to create the `Arithmetic` component. The UI for `Arithmetic` is the same as for the other `DataFrame` operation component created.

Let's add the `arithmetic` component to the `SidePlane` component:

```
import Arithmetic from './Arithmetic'
export default function SidePlanes({dataComp,
  dataComps,
  setDataComp,
```

```
    df_index,
    dfOpsType}) {

    . . . . . . . .

    if(dfOpsType === "Arithmetic") {
      return <Arithmetic
            dataComp={dataComp}
            setDataComp={setDataComp}
      />
    }

    . . . . . . . .


    }
```

The preceding code imports the `Arithmetic` component and checks whether the `dfOpsType` component is `Arithmetic`.

The following screenshot shows an example of performing an arithmetic operation on a `Data Table`:

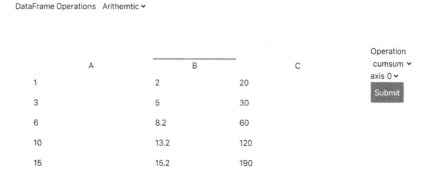

Figure 8.17 – Arithmetic operation

In this section, we discussed and implemented different `DataFrame` operations as a React component. We were able to organize some methods into a single component to prevent creating components for each of the operations.

In the next section, we will implement a `chart` component for different visualizations.

# Implementing the chart component

In this section, we'll be creating `chart` components to display common and simple charts such as a bar chart, line chart, and pie chart. We'll then implement the chart `Side Plane` to enable the setting of chart component variables.

> **Note**
>
> The code for the `Chart` and `ChartPlane` components implemented is available here: `https://github.com/PacktPublishing/ Building-Data-Driven-Applications-with-Danfo.js/ blob/main/Chapter08/src/components/ChartPlane.js`.

In the `src/components/` directory, let's create a file called `Chart.js` and the `Chart` component will be implemented with the following steps:

1.  We import our desired plotting component from `react-chartjs-2`, and then create the `Chart` component:

```
import { Bar as BarChart } from 'react-chartjs-2';
import { Line as LineChart } from "react-chartjs-2";
import { Pie as PieChart} from "react-chartjs-2";
import Draggable from 'react-draggable';

export default function Chart({labels, dataset,type}) {
  let data = {
    labels: labels,
    datasets: [{
      backgroundColor: [
      . . . . . . . .
      ],
      borderColor: [
      . . . . . . . .
      ],
      borderWidth:1,
      data: dataset,
    }]
  };
```

In the preceding code, the `Chart` component takes in the following props: `labels`, `dataset`, and `type`. `labels` signifies the column names, `dataset` represents the `dataComp` values, and `type` represents the type of chart we want to plot.

In the `Chart` component, we create a variable named `data`, which is an object formatted in the way that is required by `react-chartjs-2` for plotting our desired chart.

2.  We create a set of conditional rendering here, as we want to render a specific type of chart, based on the `prop` type passed into the `Chart` component:

```
if(type==="BarChart"){
    return(
        <Draggable>
            <div className="max-w-md">
                <BarChart data={data} options={options}
width="100" height="100" />
            </div>
        </Draggable>
    )
}
. . . . . . .
```

We check the type of chart to render. If it's a bar chart, we make a call to the `BarChart` component from `react-chartjs-2` and pass in the necessary props. The `BarChart` component is wrapped in a `Draggable` component to make the chart component being rendered draggable. The preceding code is applicable to rendering all other `Chart` components, such as `LineChart` and `PieChart` from `react-chartjs-2`.

To dive deep into `react-chartjs-2`, you can check out the documentation here: `https://github.com/reactchartjs/react-chartjs-2`.

## Implementing the ChartPlane component

We've created the `chart` component, so let's now create the chart `Side Plane`. In the `components/` folder, let's create a file called `ChartPlane.js` with the following steps:

1.  We create a `ChartPlane` component:

```
export default function ChartPlane({setChartComp,
dataComp, chartType}) {
```

```
const df = dataComp.df
const compCols = dataComp.columns
let x;
let y;
if( compCols[0] === "index") {
    x = compCols
    y = dataComp.values[0].map((val, index)=> {
        if(typeof val != "string") {
            return compCols[index]
        }
    })
} else {
    x = df.columns
    const dtypes = df.dtypes
    y = dtypes.map((val, i)=>{
        if(val != "string") {
            return x[i]
        }
    })
}
```

In the preceding code, we create a `ChartPlane` component accepting the following props:

a) `SetChartComp`: The function to update the `chartComp` state

b) `dataComp`: The current `DataTable` component to generate a chart from

c) `chartType`: The type of chart we want to generate

First, in the component, we obtain the list of possible x-axis variables and store them in the x variable. These x-axis variables can be either a column name with `String` or number dtypes.

Since we are plotting the y axis against the x axis, it is compulsory for our y axis (the y variable) to be an integer. Therefore, we check that the column of a `DataFrame` is not a string, and if not we add the column to the list of y-axis variables.

> **Note**
> This is flexible. Sometimes a chart can be flipped such that the y axis is actually the labels and the x axis contains the data.

2.  We create the UI for the ChartPlane component. Depending on how we've designed the UI for other components, the x and y variables are used to create an input field with which the user can select the *x*-axis label and the *y*-axis label:

```
<select ref={xRef} className="border">
  {
      x.map((val, index)=> {
          return <option value={val} key={index} >{val}</
option>
      })
  }
</select>
<select ref={yRef} className="border">
  {
      y.map((val, index) => {
          return <option value={val} key={index}>{val}</
option>
      })
  }
</select>
```

This UI also contains a button that triggers a function called handleChart, which updates the chart component:

```
<button onClick={()=>handleChart()} className="bg-
blue-500 p-2 text-white rounded-sm">generate Chart</
button>
```

3.  We create a function called handleChart, which obtains the value of the *x*-axis and *y*-axis input fields and uses them to create the respective charts as requested:

```
const handleChart = () => {
  const xVal = xRef.current.value
  const yVal = yRef.current.value
  const labels = xVal === "index" ? df.index : df[xVal].
values
  const data = yVal === "index" ? df.index : df[yVal].
values
    setChartComp((prev) => {
      const newChart = prev.slice()
      const key = newChart.length + 1
```

```
        const dict = {
            labels: labels,
            data: data,
            key: "chart" + key,
            type: chartType
        }
        newChart.push(dict)
        return newChart
    })
}
```

xVal and yVal are the values of the input field for the *x* axis and *y* axis. The labels and data variable are created, to contain the value of the respective columns from xVal and yVal. The label and data are then used to update the chartComp state.

## Implementing the ChartViz component

The preceding steps are used to create the chart Side Plane, but for now, we can't see the updated chartComp component. To view the chart, let's create a component to manage all chart component to be displayed.

> **Note**
>
> The code for ChartViz to be implemented is available here: https://
> github.com/PacktPublishing/Building-Data-Driven-
> Applications-with-Danfo.js/blob/main/Chapter08/
> src/components/ChartsViz.js.

Let's create a file in the components/ folder named ChartViz.js. Add the following code to the file:

```
import React from 'react'
import Chart from './Chart'


export default function ChartsViz({chartComp, setChartComp}) {
    return (
        <div>
            {
                chartComp.map((chart) => {
```

```
            return(
                <>
                <Chart
                    labels={chart.labels}
                    dataset={chart.data}
                    type={chart.type}
                />
                </>
                )
            })
            }
        </div>
    )
}
```

In the preceding code, we import our `chart` component and then create a `ChartViz` component containing the following `chartComp` and `setChartComp` props. We loop through the `chartComp` state and pass each of the state values to the `chart` component as a prop.

## Integrating ChartViz and ChartPlane into App.js

Now we are done with all the necessary parts of the `chart` component. Let's update our `App.js` component to activate the `chart` component based on the following steps:

1.  We import `ChartViz` and `ChartPlane` into `App.js`:

    ```
    import ChartsViz from './components/ChartsViz'
    import ChartPlane from './components/ChartPlane'
    ```

2.  We need to create some state to manage the type of chart we want, and the `chart` component:

    ```
    const [chartType, setChartType] = useState()
    const [chartComp, setChartComp] = useState([])
    const charts = ["BarChart", "LineChart", "PieChart"]
    ```

    In the preceding code, we also create an array variable to store a list of charts that we want to display in our `Navbar`.

3.  We create a function to update the `chartType` component and the `Side Plane` component whenever a chart is created:

```
const handleChart = (e) => { // step 4
    const value = e.target.innerHTML
    setChartType(value)
    setSidePlane("chart")
}
```

In the `handleChart` function, we obtain the target value, which is the chart type selected by the user. This value is used to update the `chartType` component and also, we notify the `Side Plane` to display a chart `Side Plane` by updating the `showSidePlane` state with the `chart` string.

4.  We loop the `charts` variable in the nav field and display them as buttons:

```
. . . . . .
{
    charts.map((chart, i) => {
        return <button disabled={dataComp.length > 0 ? false
: true}
        className={classes}
        onClick={handleChart}
        >
            {chart}
        </button>
    })
}
. . . . . .
```

In the preceding code, we loop through the `charts` array and create a button for each of the values in the array. We disabled the button by checking that the `dataComp` state is not empty, that is, whether no file has been uploaded.

5.  We call the `ChartViz` component and pass in the necessary props:

```
{(chartComp.length > 0) &&
    <ChartsViz
    chartComp={chartComp}
```

```
        setChartComp={setChartComp}
    />
}
```

We check that the `chartComp` state is not empty. If it's not, we make a call to the `ChartViz` component and then display the charts created.

6.  We then add the `ChartPlane` component:

```
<div className="border-2 w-1/3">
    <ChartPlane
        dataComp={dataComp[compIndex]}
        setChartComp={setChartComp}
        chartType={chartType}
    />
</div>
```

If the `showSide Plane` chart is a value chart, the `ChartPlane` component is displayed in the `Side Plane`.

The following screenshot shows the chart update by plotting a bar chart, line chart, and pie chart on the available `Data Table`:

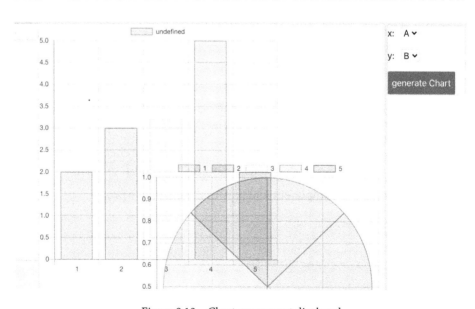

Figure 8.18 – Chart component displayed

In this section, we implemented `ChartComponent` and `ChartPlane`. We made use of `React-chart-js` to ease the development of each chart component.

## Summary

In this chapter, we saw how to create a no-code environment, where you can just upload your data and then get started with handling and doing data analysis immediately. We also saw how to convert each `DataFrame` method in Danfo.js into a React component. This gives the capability to convert all Danfo.js methods into React components, hence creating a React component library for Danfo.js.

Also, we saw how to design the flow for the app and how to manage the state in React. Even if some of the states created are redundant, this is an opportunity for you to contribute and update the app to make it robust. If you can update the app to make it possible to delete, update, and save every operation being done, this will make the app robust and even ready for production.

In the next chapter, we will be introduced to machine learning. The chapter will cover the fundamental idea behind machine learning in the simplest form possible.

# 9
# Basics of Machine Learning

The **machine learning (ML)** field is growing bigger every day, with tons of research being done, and various types of smart/intelligent applications being built with ML algorithms. The field is gaining more interest and more people are fascinated to know how it works and how to make use of it.

In this chapter, we will try to get a basic understanding of ML why and how it works, and also see various forms of its application to real-life situations.

In this chapter, we'll look at the following topics to understand the basics of ML:

- Introduction to machine learning
- Why machine learning works
- Machine learning problems/tasks
- Machine learning in JavaScript
- Applications of machine learning
- Resources to understand machine learning in depth

# Technical requirements

This chapter introduces ML in a simple form, hence it requires no prior knowledge.

# Introduction to machine learning

In this section, we will introduce ML by using a simple analogy that might serve as common ground to establish our explanation. We will also see why and how ML works.

We will start the section by using an information transfer system as a simple analogy for ML.

## A simple analogy of a machine learning system

I remember a time I was in a *Twitter* Space involving a discussion about ML and some other cool topics. I was told to give a brief introduction to ML for those who were interested but didn't fully get the gist.

The majority of people in this Twitter Space were software engineers with no previous knowledge of math, statistics, or any topic related to ML, and I came across instances where people failed to understand the terminology of the topic due to the addition of some technical terms.

This section aims to explain ML by avoiding too many technical terms and finding a common ground through which ML can be explained.

Using an information transfer system, such as a phone, information is taken from a source then encoded into a digital signal and transferred via a transfer channel to the receiver, which decodes the signal into the source input, which can be a voice, an image, and so on.

The following diagram shows the full concept of information transfer:

Figure 9.1 – Information transfer

The preceding definition is for an information transfer system whose sender and receiver are at different endpoints, but for systems such as a megaphone, the input voice is encoded into a digital signal, which is then decoded and amplified at the output.

The following shows a diagram of a megaphone:

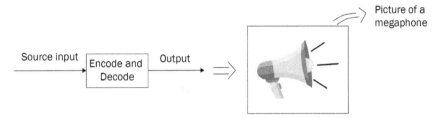

Figure 9.2 – Simple information transfer system

Using the preceding paragraph, we can establish an overview of ML. In the preceding paragraph, we mentioned some specific keywords, which are encoded and decoded.

In an information transfer system, a large body of information – either voice or image – is encoded or compressed into a digital signal at the source end and then decoded back to the source information at the output end.

The same thing described in the preceding paragraph goes for a ML system – a large body of information is encoded or compressed into *Forms of representation* (mind the highlighted words) and then decoded into conceptual or intelligent or decisional output.

Note the terms *digital signal* and *form of representation* in the two preceding paragraphs. In an information transfer system, there is some information theory that is responsible for converting any input no matter the form (any type of image, any type of sound/voice) into a digital signal.

But in ML, we have some forms of theories and algorithms. These algorithms do not just process input information and give an output. First, a sample of information is obtained. This information is processed and used to build a form of representation that summarizes the whole information and maps it to a decision output.

This form of representation is used to build the final ML system, which takes an input source, compares it to the form of representation, and outputs a decoded decision (intelligent output) that matches the comparison between the source input and the form of representation.

The following diagram shows the conceptual illustration of the two preceding paragraphs:

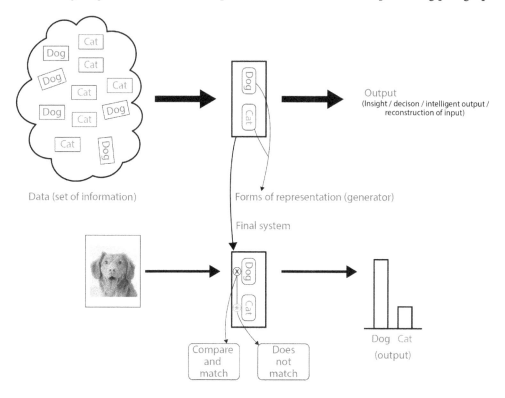

Figure 9.3 – Conceptual illustration of machine learning

There are some key things about ML to note from the preceding paragraphs, as follows:

- First, we generate a form of representation from a large set of information. Also, note that the process of generating a form of representation from a set of information is called **Training**.

- The form of representation generated is then used to create the final ML system, and this process is called **Inference**.

In the next sub-section, we will see how this form of representation is generated, the whole idea behind generating a form of representation from a large set of information, and then using this representation to build the final ML system.

# Why machine learning works

During our illustration of training the ML model, we talked about generating a form of representation that is used to build our ML system. One thing to note is that this information or the data used to generate the form of representation is the data representation of our future source of information. Why do we need the data representation of our future source of information? In this sub-section, we will look into this and see how it helps in creating ML models.

Let's assume we are told to carry out research about product interest in a particular community. Imagine this community consists of a large number of people and we are only able to reach a few people – let's say 50% of the community population.

The idea is that, from the information obtained from 50% of the population, we should be able to generalize for the remaining 50% of the population. We assume this because a set of people from the same community or population are assumed to share quite a number of the same attributes and beliefs. Therefore, if we use information obtained from 50% of the individuals of this population to train our model, our model should be able to distinguish any information that comes from any individual belonging to the same population.

In the worst-case scenario, there might be some outliers in the population – folks who don't have the same beliefs as the rest of the people, or the individual information we obtained from 50% of the population might not capture the attributes of the other 50% of the population. In this case, the model will fail if this information is passed into the model.

The preceding paragraph shows why in ML by default, the more data there is, the better the ML model. The following diagram shows a sample distribution (our 50%-of-individuals information) and the population itself:

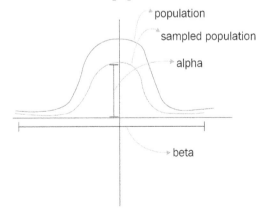

Figure 9.4 – Population distribution

From *Figure 9.3*, when we say we are training in ML, we mean the ML algorithm is learning the parameters that control and generalize the population from our sampled population (this parameter is the form of representation in this case). We have two parameters, beta and alpha, and the goal of our training is for the model to obtain the best value from these parameters that control the population.

Let's see a more concrete example: we want to create an app that assigns a particular product to dogs alone. But as you know, we have different breeds of dogs and dogs also have some facial features that are similar to those of cats.

To create the ML model for this app, we sample a number of dog images, but this sample does not capture all breeds of dogs. The goal of an ML model is to capture the unique attributes/parameters of dogs alone from the data given to it (these unique attributes/parameters are the forms of representation).

If the model is good, it should be able to tell whether an input image is a dog or not. Then how do we measure how good the model is? The following methods are used to achieve that:

- **Objective functions**
- **Evaluation metrics**

In the following sub-sections, we will see how these methods work.

## Objective functions

We've seen how to sample data from a large population and use that to train our model and hope that the model generalizes well. During the course of training, we want to measure how close our model is to our goals, and to do that we create an objective function. Some call this function by a different name, such as the loss function or error rate. The lower the score returned by this function, the better our model is.

To classify whether an image is a dog or not, we have our dataset containing images of dogs and cats, for example. This dataset also contains labels. Each image in the dataset has a label, which informs us whether the image in the dataset is a dog image or a cat image.

The following shows an example of what the dataset will look like:

| data(images) | Labels |
| --- | --- |
| /001.jpeg | dog |
| /002.jpeg | cat |
| /003.jpeg | dog |
| /004.jpeg | dog |
| /005.jpeg | cat |

Figure 9.5 – Dataset sample

During training, each of the images in the data, as shown in *Figure 9.5*, is passed as input into the model, and the model predicts the label. The label predicted by the model is compared with the labels shown in *Figure 9.4* by the objective function. We keep training the model until the model predicts the true label of each image in the dataset.

The model might be able to classify all the images in the dataset properly based on the objective function, but that does not mean the model generalizes well, that is, the model might be able to classify some dog images correctly during training, but when given images are not available in the dataset, as shown in *Figure 9.4*, the model misclassifies the images. This leads us to the second method.

## Evaluation metrics

We've trained our model and it is giving us a very low loss score, which is good, but we will need to be sure of whether the model has captured the attributes for the whole population or just for the sampled population of the dataset used for training. What am I saying? It is possible for the model to perform well while training but actually be bad if we are to test it on other images containing dogs and cats.

To check whether the model is good and has captured attributes unique to each population of dogs and cats, we test the model on a set of datasets, which is also a sample from the same population as the one used for training. If the model is able to give us a better score, then the model is good; if the score is bad compared to that from the objective function, then the model is bad. This process is called the evaluation process and we use different metrics to measure the performance of the model. The metrics are called **Evaluation metrics**.

The following diagram shows the pipeline of a ML model:

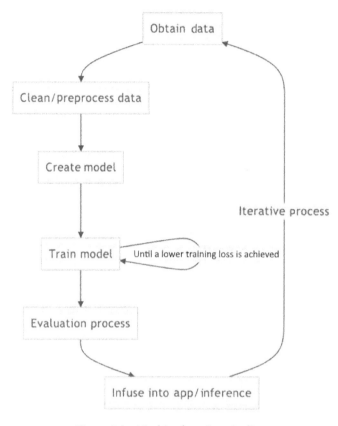

Figure 9.6 – Machine learning pipeline

In this section, we discussed ML-based information transfer. We saw how an ML model works and the basic workflow of an ML model. In the next section, we will talk about the grouping of ML tasks into different categories.

# Machine learning problems/tasks

ML problems or tasks such as classification problems can be categorized into different groups based on how the model learns.

In this section, we will look into two of the most popular categories of ML problems:

- **Supervised learning**
- **Unsupervised learning**

First, we will look into supervised learning.

## Supervised learning

In this category, the model learns under supervision. By supervision, we mean the model knows whether it is doing well based on the provided label. While training, we provide the model with a dataset containing a set of labels, which are used to correct and improve the model. With this, we can measure how well the model performs.

The following ML problems/tasks belong to this category:

- **Classification problems**: In this type of problem, the model is made to classify an input to a set of discrete categories, such as classifying whether the image is a dog or a cat.
- **Regression problems**: This involves the model mapping the input to a set of continuous values. For example, creating a model to predict the price of a house given some features of the house.

The following diagram shows what classification is:

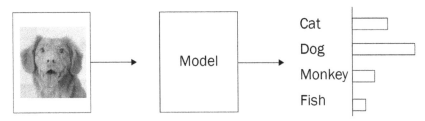

Figure 9.7 – Classification problem

The following diagram shows the illustration of a regression problem:

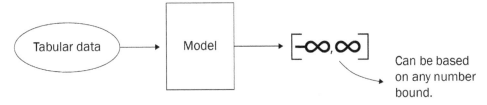

Figure 9.8 – Regression problem

In summary, a supervised learning algorithm is used for problems in which there is a label provided with the dataset, where the label is used to measure the performance of the model. Sometimes we have data, but we don't have a ground truth that is a label to scale how the model performs. This leads us to unsupervised learning.

## Unsupervised learning

When we don't have labels, but we have our data, what can we do? The best thing to do is to draw insight from the data.

Remember the population example at the beginning of the section? Let's say we sample some set of entities from a population but have no prior knowledge of their behavior. The best thing is to study them for some time so we can understand their likes and dislikes and find out what makes them unique.

Through this observation, we can group the population into categories based on their beliefs, occupation, food tastes, and so on.

The following ML problems belong to the unsupervised learning category:

- **Clustering problems**: Clustering problems involve revealing some hidden attribute in a dataset (our sampled population) and then grouping each entity in the population based on this attribute.

- **Association problems**: This involves the discovery association rule in a population. It involves knowing whether people that engage in one activity also engage in another activity.

The main gist of this is that we want to obtain a hidden insight from this dataset as shown in the following diagram:

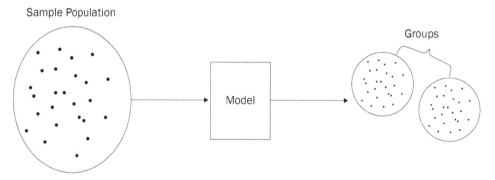

Figure 9.9 – Unsupervised learning (clustering example)

In this section, we looked into some categories of ML problems, and we also saw a scenario in which each ML problem category is important and the kind of tasks they are meant to be used for.

In the next section, we will talk about making ML more accessible.

# Machine learning in JavaScript

The web is the most accessible platform and JavaScript is the language used across the web, hence ML in JavaScript gives us more control and accessibility. In the *Why you need Danfo.js* section of *Chapter 3, Getting Started with Danfo.js*, we talked about the importance of bringing ML the web. We also talked about how browsers' computational power is increasing and how this is a benefit to JavaScript for ML.

In this section, I will list some open source tools for ML tasks in the browser:

- **TensorFlow.js (tfjs)** (`https://github.com/tensorflow/tfjs`): A WebGL accelerated JavaScript library for training and deploying ML models.

- **datacook** (`https://github.com/imgcook/datacook`): A JavaScript framework for feature engineering on datasets.

- **Nlp.js** (`https://github.com/axa-group/nlp.js`): A JavaScript framework for NLP tasks such as sentiment analysis, automatic language identity, entity extraction, and so on.

- **Natural** (`https://github.com/NaturalNode/natural`): Also, NLP, it covers almost all the necessary algorithms for NLP tasks.

- **Pipcook** (`https://github.com/alibaba/pipcook`): A machine learning platform for web developers.

- **Jimp** (`https://github.com/oliver-moran/jimp`): An image processing library written entirely in JavaScript.

- **Brain.js** (`https://github.com/BrainJS/brain.js`): A GPU accelerated neural network in JavaScript for the browser and Node.js.

The preceding tools are the most popular and have recent updates. By using these tools, you can integrate ML into your next web app.

In the next section, we will look into some applications of ML in the real world.

# Applications of machine learning

ML is transforming software development and is also making things more *automatic*, *self-driving*, and *self-operating*. In this section, we will look into some examples of ML applications.

The following are examples of machine learning applications:

- **Machine translation**: ML enables us to build software that easily translates a language to another language.

- **Games**: With some advanced ML algorithms, some software is becoming better at playing more complicated games such as the game Go and beating world champions at what they do best. For example, here's a video about **AlphaGo**: `https://www.youtube.com/watch?v=WXuK6gekU1Y`.

- **Vision**: Machines are getting better at seeing and providing meaning for what they see.

  a) **Self-driving cars**: ML is helping to create fully self-driving cars.

  b) **Tesla demonstration of a self-driving car**: `https://www.youtube.com/watch?v=VG68SKoG7vE`

- **Recommendation engines**: ML algorithms are improving recommendation engines and hooking customers.

  How *Netflix* is using ML for personalized recommendations: `https://netflixtechblog.com/artwork-personalization-c589f074ad76`

- **Art**: ML is used to generate artworks, new stories, new paintings, and new images.

  a) Here is a website that generates images of people that never existed: `https://thispersondoesnotexist.com/`.

  b) A generated art gallery: `https://www.artaigallery.com/`.

  c) Architectural design with ML: `https://span-arch.org/`

In this section, we saw a few examples of how ML is being used for different purposes. In the next section, we will provide some materials to better understand ML.

# Resources to understand machine learning in depth

In this section, we will provide resources to better understand ML in depth and get better at creating software making use of ML algorithms.

The following are resources that can be used to understand ML:

- **fastai** (`https://www.fast.ai/`): This community provides courses, frameworks, and books for ML practitioners.
- **Cs231n** (`http://cs231n.stanford.edu/`): This course gives the fundamentals of **deep learning** and introduces you to computer vision.
- **Hugging Face**: Hugging Face provides the best framework for **natural language processing** and different transformer models. It also has a course (`https://huggingface.co/course/chapter1`) that provides a full detail of transformer models and deployment.
- **Andrew Ng course**: A ML course on *YouTube* that also provides full ML details.

There are tons of available materials to learn about ML online. Just follow one path and follow it to the end and avoid jumping from one lecture to another.

# Summary

In this chapter, we looked into ML using the concept of information transfer. We then looked into how and why it works. We also talked about the idea of sampling from a population to understand the population.

We talked about different categories of ML problems and also discussed some tools needed for ML for the web platform, and we also showed some examples of real-world applications of ML.

The intention of this chapter was to get the whole idea of ML to aid understanding during personal learning.

In the next chapter, we will introduce **TensorFlow.js**. TensorFlow.js is useful when integrating ML into your web apps.

# 10
# Introduction to TensorFlow.js

In the previous chapter, you were introduced to the basics of **machine learning** (**ML**), and you learned some theoretical foundations that are required in order to build and use ML models.

In this chapter, we'll introduce you to an efficient and popular ML library in JavaScript called TensorFlow.js. By the end of this chapter, you'll know how to install and use TensorFlow.js, how to create tensors, how to operate on tensors using the Core **application programming interface** (**API**), as well as how to build a regression model using TensorFlow.js's Layer API.

In this chapter, we will cover the following topics:

- What is TensorFlow.js?
- Installing and using TensorFlow.js
- Tensors and basic operations on tensors
- Building a simple regression model with TensorFlow.js

# Technical requirements

To follow along in this chapter, you should have these tools or resources:

- A modern browser such as Chrome, Safari, Opera, or Firefox.

- Node.js installed on your system

- A stable internet connection for downloading packages and datasets

- The code for this chapter is available and can be cloned from GitHub at `https://github.com/PacktPublishing/Building-Data-Driven-Applications-with-Danfo.js/tree/main/Chapter10`

# What is TensorFlow.js?

**TensorFlow.js** (**tfjs**) is a JavaScript library for creating, training, and deploying ML models in the browser or in Node.js. It was created at Google by Nikhil Thorat and Daniel Smilkov and was initially called Deeplearn.js, before being merged into the TensorFlow team in 2018 and renamed as TensorFlow.js.

TensorFlow.js provides two main layers, outlined as follows:

- **CoreAPI**: This is the low-level API that deals directly with tensors—the core data structure of TensorFlow.js.

- **LayerAPI**: A high-level layer built on top of the CoreAPI layer for easily building ML models.

In later sections, *Tensors and basic operations on tensors* and *Building a simple regression model with TensorFlow.js*, you will learn more details about the CoreAPI and LayerAPI layers.

With TensorFlow.js, you can do the following:

- Perform hardware-accelerated mathematical operations

- Develop ML models for the browser or Node.js

- Retrain existing ML models using **transfer learning** (**TL**)

- Reuse existing ML models trained with Python

In this chapter, we will cover performing hardware-accelerated mathematical operations and developing ML models with TensorFlow.js. If you want to learn about the last two use cases—retraining and reusing ML models—then the official TensorFlow.js documentation (`https://www.tensorflow.org/js/guide`) is a great place to start.

Now we have the introduction out of the way, in the next section, we'll show you how to install and use TensorFlow.js in both the browser and Node.js environment.

# Installing and using TensorFlow.js

As we mentioned earlier, TensorFlow.js can be installed and run in both the browser and Node.js environment. In the following paragraphs, we'll show you how to achieve this, starting off with the browser.

## Setting up TensorFlow.js in the browser

There are two ways of installing TensorFlow.js in the browser. These are outlined here:

- Via script tags
- Using package managers such as **Node Package Manager** (**npm**) or **Yarn**

### Installing via script tags

Installing TensorFlow.js via a `script` tag is easy. Just place the `script` tag in the header file of your **HyperText Markup Language** (**HTML**) file, as shown in the following code snippet:

```
<script src="https://cdn.jsdelivr.net/npm/@tensorflow/
tfjs@3.6.0/dist/tf.min.js"></script>
```

To confirm that TensorFlow.js is installed, open the HTML file in the browser, and check the network tabs. You should see the name `tf.min.js` and a status code of `200`, as shown in the following screenshot:

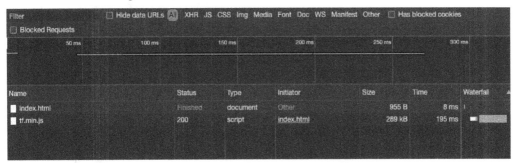

Figure 10.1 – Network tab showing the successful installation of tfjs

You can add a simple script in the body of your HTML file to confirm the successful installation of `tfjs`. In a `script` section of your HTML file, add the following code:

```
. . .
<script>
```

```
            tf.ready().then(()=>{
                console.log("Tensorflow.js loaded successfully!");
            })
    </script>
    ...
```

The preceding code snippet logs the text `Tensorflow.js loaded successfully!` to the browser console, as soon as TensorFlow.js is loaded and ready on the page. To see the output, open the HTML file in the browser and check the console output. You should see an output result, as shown in the following screenshot:

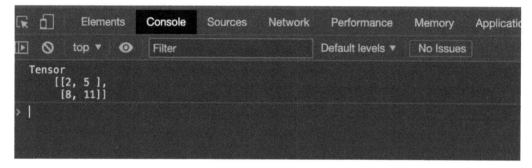

Figure 10.2 – Tensor output from add operation

Next, let's see how to install `tfjs` via package managers.

## Installing via package managers

You can install `tfjs` via package managers such as `npm` or `yarn`. This is useful when you need to use `tfjs` in client-side projects such as React and Vue projects.

To install with `npm`, run the following command in your **command-line interface (CLI)**:

```
npm install @tensorflow/tfjs
```

And to install with `yarn`, just run the following command in the CLI as well:

```
yarn add @tensorflow/tfjs
```

> **Note**
>
> Before you can successfully install packages using npm or yarn via the CLI,
> you must have either of them installed in your system—preferably globally. If
> you have Node.js installed, then you already have npm. To install yarn, you
> can follow the steps here: https://classic.yarnpkg.com/en/
> docs/install/#mac-stable.

On successful installation, you can import and use tfjs, as shown in the following
code snippet:

```
import * as tf from '@tensorflow/tfjs';
const x = tf.tensor2d([1, 2, 3, 4], [2, 2]);
const y = tf.tensor2d([1, 3, 5, 7], [2, 2]);
const sum = x.add(y)
 sum.print()
```

Running the preceding code snippet will produce the following output in the console:

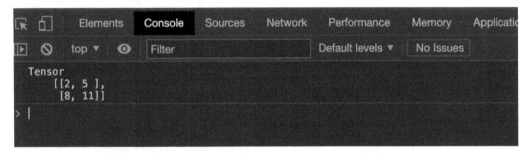

Figure 10.3 – Output from testing tfjs installed with package managers

By following the steps in the preceding code blocks, you should be able to install and use
tfjs in the browser or client-side frameworks. In the next section, we'll show you how to
install tfjs in a Node.js environment.

# Installing TensorFlow.js in Node.js

Installing tfjs in Node.js is quite simple, but first, ensure you have Node.js, npm, or
yarn installed on your system.

TensorFlow.js in Node.js has three options, and the choice of installation will
depend on your system specification. In the following sub-sections, we'll show
you these three options.

## Installing TensorFlow.js with native C++ bindings

The `@tensorflow/tfjs-node` (`https://www.npmjs.com/package/@tensorflow/tfjs-node`) version of `tfjs` connects directly to TensorFlow's native C++ bindings. This makes it fast, as well as giving it a close performance with the Python version of TensorFlow. This means that both `tfjs-node` and `tf.keras` use the same C++ bindings under the hood.

To install `tfjs-node`, just run the following command via the CLI:

```
npm install @tensorflow/tfjs-node
```

Or, if using `yarn`, run the following command via the CLI:

```
yarn add @tensorflow/tfjs-node
```

## Installing TensorFlow.js with GPU support

The `@tensorflow/tfjs-node-gpu` version of `tfjs` provides support for running operations on **graphics processing unit** (**GPU**)-enabled hardware. Operations run using `tfjs-node-gpu` are generally faster than that of `tfjs-node` as operations can be easily vectorized.

To install `tfjs-node-gpu`, just run the following command via the CLI:

```
npm install @tensorflow/tfjs-node-gpu
```

Or, if you're using `yarn`, run the following command via the CLI:

```
yarn add @tensorflow/tfjs-node-gpu
```

## Installing plain TensorFlow.js

The `@tensorflow/tfjs` version is the pure JavaScript version of `tfjs`. It is the slowest in terms of performance and should rarely be used.

To install this version, just run the following command via the CLI:

```
npm install @tensorflow/tfjs
```

Or, if you're using `yarn`, run the following command via the CLI:

```
yarn add @tensorflow/tfjs
```

If you followed the preceding steps, then you should have at least one of the versions of tfjs installed. You can test for successful installation using the following code example:

```
const tf = require('@tensorflow/tfjs-node')
// const tf = require('@tensorflow/tfjs-node-gpu') GPU version
// const tf = require('@tensorflow/tfjs') Pure JS version
const xs = tf.randomNormal([100, 10])
const ys = tf.randomNormal([100, 1])
const sum = xs.add(ys)
const xsSum = xs.sum()
const xsMean = xs.mean()


console.log("Sum of xs and ys")
sum.print()
console.log("Sum of xs")
xsSum.print()
console.log("Mean of xs")
xsMean.print()
```

> **Note**
>
> We call the print() function on a tensor when we want to see the underlying data. If we use the default console.log, we get the Tensor object instead.

Running the preceding code should output the following in the console:

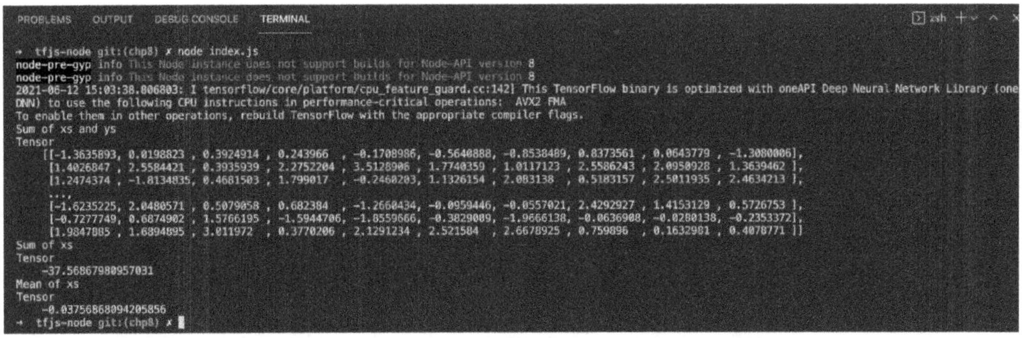

Figure 10.4 – Output from testing tfjs installed in Node.js

Now you have successfully installed t f j s in your project, in the next section, we'll introduce you to the core data structure of t f j s—tensors.

# Tensors and basic operations on tensors

A tensor is a basic data structure in t f j s. You can think of tensors as a generalization of vectors, matrices, or high-dimensional arrays. The **CoreAPI**, which we introduced in the *What is TensorFlow.js?* section, exposes different functions for creating and working with tensors.

The following screenshot shows a simple comparison between scalars, vectors, and a matrix with a tensor:

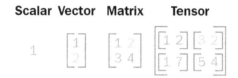

Figure 10.5 – Comparison between simple n-dimensional arrays and a tensor

> **Tip**
>
> A matrix is a grid of m  x  n numbers, where m represents the number of rows and n represents the number of columns. A matrix can be of one or more dimensions, and matrixes of the same shape support direct mathematical operations on each other.
>
> A vector, on the other hand, is a one-dimensional matrix with shape (1, 1); that is, it has a single row and column—for example, [2, 3], [3, 1, 4].

We mentioned earlier that a tensor is more of a generalized matrix—that is, it extends the concept of a matrix. Tensors can be described by their rank. A rank is similar to the idea of a shape but is represented by a single number as opposed to a shape. In the following list, we see the different types of tensor ranks with examples:

- A tensor of rank 0 is a scalar—for example, 1, 20, or 100.

- A tensor with rank 1 is a vector—for example, [1, 20] or [20, 100, 23.6].

- A tensor with a rank of 2 is a matrix—for example, [[1, 3, 6], [2.3, 5, 7]].

Note that we can have tensors of rank 4 or more, and these are called higher dimension tensors and can be difficult to visualize. See the following screenshot for a better understanding of tensors:

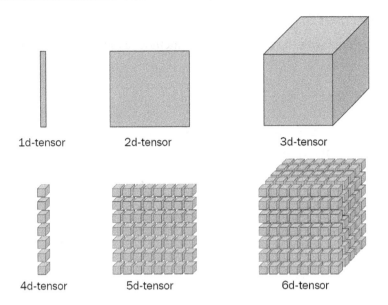

Figure 10.6 – Comparison of tensors with different ranks

Besides the rank, tensors have other properties such as `dtype`, `data`, `axis`, and `shape`. These are described in more detail here:

- The `dtype` property (data type) is the type of data a tensor holds—for example, a rank 1 tensor with the following data [2.5, 3.8] has a dtype of `float32`. By default, numeric tensors have a dtype of `float32`, but this can be changed during creation. TensorFlow.js supports `float32`, `int32`, `bool`, `complex64`, and `string` data types.

- The `data` property is the content of a tensor. This is usually stored as an array.

- The `axis` property is the particular dimension of a tensor—for example, an *m x n* tensor has an axis of *m* or *n*. The axis can be used to specify which dimension an operation is performed on.

- The `shape` property is the dimension of the tensor. Think of the shape as the number of elements across each axis of a tensor.

Now you have a basic understanding of what tensors are, in the next sub-section, we'll show you how to create tensors and perform some basic operations on or with them.

## Creating tensors

A tensor can be created with the `tf.tensor()` method, as shown in the following code snippet:

```
const tf = require('@tensorflow/tfjs-node')

const tvector = tf.tensor([1, 2, 3, 4]);
console.log(tvector)
//output
Tensor {
  kept: false,
  isDisposedInternal: false,
  shape: [ 4 ],
  dtype: 'float32',
  size: 4,
  strides: [],
  dataId: {},
  id: 0,
  rankType: '1'
}
```

In the preceding code snippet, we pass a flat array (vector) to the `tf.tensor()` method to create a `tfjs` tensor. After creating this, we now have access to different properties and functions that can be used to manipulate or transform the tensor.

One such property is the `shape` property, which we can call as shown in the following code snippet:

```
console.log('shape:', tvector.shape);
//outputs: shape: [ 4 ]
```

Notice that when you log the tensor with `console.log`, you get a tensor object. If you need to see the underlying tensor array, you can call the `print()` function on the tensor, as shown in the following code snippet:

```
tvector.print();
//outputs
Tensor
    [1, 2, 3, 4]
```

If you need access to the underlying data of a tensor, you can call the `array()` or `arraySync()` method. The difference between the two is that `array()` runs asynchronously and returns a promise that resolves to the underlying array, while `arraySync()` runs synchronously. You can see an example of this here:

```
const tvectorArray = tvector.array()
const tvectorArraySync = tvector.arraySync()
console.log(tvectorArray)
console.log(tvectorArraySync)
//outputs
Promise { <pending> }
[ 1, 2, 3, 4 ]
```

You can also create tensors by specifying a `shape` parameter. For example, in the following code snippet, we create a 2 x 2 (**two-dimensional (2D)**) tensor from a flat array:

```
const ts = tf.tensor([1, 2, 3, 4], [2, 2]);
console.log('shape:', ts.shape);
ts.print();
//outputs
shape: [ 2, 2 ]
Tensor
    [[1, 2],
     [3, 4]]
```

Or, we can create a 1 x 4 (**one-dimensional (1D)**) tensor, as shown in the following code snippet:

```
const ts = tf.tensor([1, 2, 3, 4], [1, 4]);
console.log('shape:', ts.shape);
ts.print();
//outputs
shape: [ 1, 4 ]
Tensor
    [[1, 2, 3, 4],]
```

Note, however, that the shapes must match the number of elements—for example, you cannot create a 2 x 5-dimensional tensor from a flat array with four elements. The following code will throw a shape error:

```
const ts = tf.tensor([1, 2, 3, 4], [2, 5]);
```

The output is shown here:

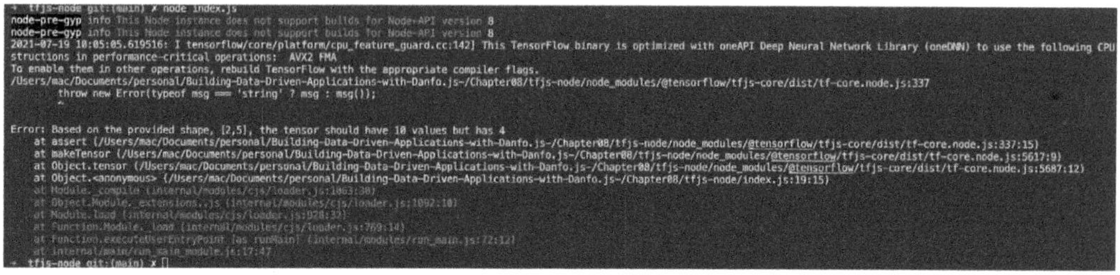

Figure 10.7 – Error thrown from shape mismatch

Tfjs explicitly provides functions for creating 1D, 2D, **three-dimensional (3D)**, **four-dimensional (4D)**, **five-dimensional (5D)**, and **six-dimensional (6D)** tensors. You can use this instead of specifying a shape parameter. You can read more about creating tensors from the official tfjs API here: https://js.tensorflow.org/api/latest/#Tensors-Creation.

By default, a tensor has a dtype property of float32, so any and every tensor you create will have a dtype of float32. If this is not the desired dtype, you can specify the type on tensor creation, as we demonstrate in the following code snippet:

```
const tsInt = tf.tensor([1, 2, 3, 4], [1, 4], 'int32');
console.log('dtype:', tsInt.dtype);
//outputs
dtype: int32
```

Now you know how to create a tensor, we are going to move on to operating on tensors.

## Operating on tensors

Tensors, as we said earlier, store data in grids and allow numerous operations for manipulating or transforming this data. tfjs provides many operators for linear algebra and ML.

Operations in `tfjs` are grouped into different sections. Here is an explanation of some of the common operations:

- **Arithmetic**: These operators are used to perform arithmetic computation on tensors. Tensors are immutable, so all operations always return new tensors and never modify input tensors—for example, `add()` for the addition of tensors, `sub()` for the subtraction of tensors, `mul()` for the multiplication of tensors, and `div()` for the division of tensors. See a full list with examples here: `https://js.tensorflow.org/api/3.7.0/#Operations-Arithmetic`.

- **Basic math**: These operators are used to perform basic mathematical computation on Tensors—for example, `cos()` for computing the cosine of a tensor, `sin()` for computing the sine of a tensor, `exp()` for computing the exponential of a tensor, and `log()` for computing the natural logarithm of a tensor. See a full list with examples here: `https://js.tensorflow.org/api/3.7.0/#Operations-Basic%20math`.

- **Matrices**: These operators are used for matrix operations such as dot products, norms, or transposes. You can see a full list of supported operators here: `https://js.tensorflow.org/api/3.7.0/#Operations-Matrices`.

- **Convolution**: These are operators for computing convolutions on tensors— for example, `conv1d`, which computes a 1D convolution over the input x, and `maxpool3D`, which computes a 3D max-pooling operation. See a full list here: `https://js.tensorflow.org/api/3.7.0/#Operations-Convolution`.

- **Reduction**: These are operators for computing tensor reductions—for example, `min`, `max`, `sum`, `mean`, `argMax`, and `argMin`. You can see a full list with examples here: `https://js.tensorflow.org/api/3.7.0/#Operations-Reduction`.

- **Logical**: These are operators for computing Boolean logic on tensors—for example, `equal`, `greater`, `greaterEqual`, and `less`. You can see a full list with examples here: `https://js.tensorflow.org/api/3.7.0/#Operations-Logical`.

You can see a full list of supported operations in the official API here: `https://js.tensorflow.org/api/3.7.0/#Operations`.

Now you have a basic understanding of the available tensor operators, we'll show some code examples.

## Applying arithmetic operations on tensors

We can add two tensors by directly calling the add() method on the first tensor and passing the second tensor as an argument, as illustrated in the following code snippet:

```
const tf = require('@tensorflow/tfjs-node')
const a = tf.tensor1d([1, 2, 3, 4]);
const b = tf.tensor1d([10, 20, 30, 40]);
a.add(b).print();
//outputs
Tensor
    [11, 22, 33, 44]
```

Note that you can also add or apply any operator directly by calling the operator on the tf object, as shown in the following code snippet:

```
const tf = require('@tensorflow/tfjs-node')
const a = tf.tensor1d([1, 2, 3, 4]);
const b = tf.tensor1d([10, 20, 30, 40]);
const sum = tf.add(a, b)
sum.print()
//outputs
Tensor
    [11, 22, 33, 44]
```

Using this knowledge, you can perform other arithmetic operations such as subtraction, multiplication, division, and power operations, as we demonstrate in the following code snippet:

```
const a = tf.tensor1d([1, 2, 3, 4]);
const b = tf.tensor1d([10, 20, 30, 40]);

const tfsum = tf.add(a, b)
const tfsub = tf.sub(b, a)
const tfdiv = tf.div(b, a)
const tfpow = tf.pow(b, a)
const tfmax = tf.maximum(a, b)

tfsum.print()
tfsub.print()
```

```
tfdiv.print()
tfpow.print()
tfmax.print()
//outputs
Tensor
    [11, 22, 33, 44]
Tensor
    [9, 18, 27, 36]
Tensor
    [10, 10, 10, 10]
Tensor
    [10, 400, 27000, 2560000]
Tensor
    [10, 20, 30, 40]
```

It is worth mentioning that the order of the tensors passed to the operators matter, as a change in the order causes the result to be different. For example, if we swap the order of the preceding `div` operation from `const tfsub = tf.sub(b, a)` to `const tfsub = tf.sub(a, b)`, then we get a negative result, as shown in the following output:

```
Tensor
    [-9, -18, -27, -36]
```

Note that all operations involving two tensors will only work if both tensors have the same shape. For example, the following operation would throw an invalid shape error:

```
const a = tf.tensor1d([1, 2, 3, 4]);
const b = tf.tensor1d([10, 20, 30, 40, 50]);
const tfsum = tf.add(a, b)
```

Figure 10.8 – Invalid shape error when performing an operation on tensors with a different shape

In the next sub-section, we look at some examples of basic math operations on tensors.

## Applying basic math operations on tensors

Following the example format from the previous sub-section, *Applying arithmetic operations on tensors*, we give some examples of computing math operations on tensors, as follows:

```
const tf = require('@tensorflow/tfjs-node')

const x = tf.tensor1d([-1, 2, -3, 4]);
x.abs().print();   // Computes the absolute values of the tensor
x.cos().print(); // Computes the cosine of the tensor
x.exp().print(); // Computes the exponential of the tensor
x.log().print(); // Computes the natural logarithm  of the
tensor
x.square().print(); // Computes the sqaure of the tensor
```

The output is shown here:

```
Tensor
    [1, 2, 3, 4]
Tensor
    [0.5403023, -0.4161468, -0.9899925, -0.6536436]
Tensor
    [0.3678795, 7.3890562, 0.0497871, 54.5981522]
Tensor
    [NaN, 0.6931472, NaN, 1.3862944]
Tensor
    [1, 4, 9, 16]
```

As we mentioned earlier, you can call operators directly from the `tf` object—for example, `x.cos()` becomes `tf.cos(x)`.

## Applying reduction operations on tensors

We can also apply reduction operations such as mean, min, max, argMin, and argMax to tensors. Here are some examples of mean, min, max, argMin, and argMax in the following code snippet:

```
const x = tf.tensor1d([1, 2, 3]);
x.mean().print();   // or tf.mean(x)  Returns the mean value of
the tensor
x.min().print();   // or tf.min(x) Returns the smallest value in
the tensor
x.max().print();   // or tf.max(x) Returns the largest value in
the tensor
x.argMax().print();   // or tf.argMax(x)  Returns the index of
the largest value
x.argMin().print();   // or tf.argMin(x)  Returns the index of
the smallest value
```

The output is shown here:

```
Tensor 2
Tensor 1
Tensor 3
Tensor 2
Tensor 0
```

Armed with the basic knowledge of ML, tensors, and operations that can be performed on tensors, you are now ready to build a simple ML model. In the next section of this chapter, we will consolidate all that you have learned in this section.

# Building a simple regression model with TensorFlow.js

In the previous chapter, *Chapter 9, Basics of Machine Learning*, you were introduced to the basics of ML, especially the theoretical aspect of regression and classification models. In this section, we'll show you how to create and train a regression model using tfjs **LayerAPI**. Specifically, by the end of this section, you'll have a regression model that can predict sales prices from supermarket data.

# Setting up your environment locally

Before building the regression model, you have to set up your environment locally. In this section, we'll be working in a Node.js environment. This means that we'll be using the node version of TensorFlow.js and Danfo.js.

Follow the steps here to set up your environment:

1.  In a new work directory, create a folder for your project. We will create one called `sales_predictor`, as demonstrated in the following code snippet:

    ```
    mkdir sales_predictor
    cd sales_predictor
    ```

2.  Next, open a Terminal in the folder directory and initialize a new npm project by running the following command:

    ```
    npm init
    ```

3.  Next, install the `Danfo.js` node package, as follows:

    ```
    yarn add danfojs-node
    or if using npm
    npm install danfojs-node
    ```

4.  Also from the terminal, create an `src` folder and add `train.js`, `model.js`, and `data_proc.js` files. You can create these folders/files manually from your code editor or by running the following command in the terminal:

    ```
    mkdir src &&
    cd src &&
    touch train.js &&
    touch data_proc.js &&
    touch model.js
    ```

    > **Note**
    >
    > Notice that in the preceding code snippet, we created three files (`train.js`, `data_proc.js`, and `model.js`) in the `src` folder. These files will contain code for processing data, creating a `tfjs` model, and model training, respectively.

Now you have your project and files set up, we'll move on to the data retrieval and processing step in the next section.

# Retrieving and processing the training dataset

The dataset we'll be using for model training is called the *BigMart sales dataset* (`https://www.kaggle.com/devashish0507/big-mart-sales-prediction`). It is available as a public dataset on Kaggle, which is a popular data science competition platform.

You can download the dataset directly from this chapter's code repository here: `https://github.com/PacktPublishing/Building-Data-Driven-Applications-with-Danfo.js-/blob/main/Chapter10/sales_predictor/src/dataset/Train.csv`. After successful download, create a folder called `dataset` in your project directory and copy the dataset into it.

To confirm everything is in order, your project `src` folder should have the following file structure:

```
|-data-proc.js
|-dataset
|       └── Train.csv
|-model.js
|-train.js
```

As with all data science problems, a general problem statement is made available to guide you on the problem you're solving. In terms of the BigMart sales dataset, the problem statement is as follows:

*BigMart have collected 2013 sales data for 1,559 products across 10 stores in different cities. Also, certain attributes of each product and store have been defined. The aim is to build a predictive model and find out the sales of each product at a particular store.*

From the preceding problem statement, you will notice that the aim of building this model is to help BigMart effectively predict sales of each product at a particular store. Now, sales price here means a continuous value, so as such, we have a regression problem.

Now you have access to the data and have understood the problem statement, you're going to load the dataset with `Danfo.js` and perform some data processing and cleaning.

> **Note**
>
> We have provided a separate **Danfo Notebook (Dnotebook)** file in the code repository here: `https://github.com/PacktPublishing/ Building-Data-Driven-Applications-with-Danfo.js-/ blob/main/Chapter10/sales_predictor/src/bigmart%20 sales%20notebook.json`. In the notebook, we did some data exploration and analysis of the sales dataset, a majority of which will help us in the following processing steps.

With the `data_proc.js` file opened in your code editor, follow the steps given here to process the BigMart sales dataset:

1.  First, we will import `danfojs-node`, as follows:

    ```
    const dfd = require("danfojs-node")
    ```

2.  Then, we create a function called `processData` that accepts the dataset path, as follows:

    ```
    async function processData(trainDataPath) {
        //… process code goes here
    }
    ```

3.  Next, in the body of the `processData` function, we load the dataset using the `read_csv` function and print the header, as follows:

    ```
    const salesDf = await dfd.read_csv(trainDataPath)
    salesDf.head().print()
    ```

4.  To ensure data loading works, you can pass the path of your dataset to the `processData` function, as shown in the following code snippet:

    ```
    processData("./dataset/train.csv")
    ```

5.  And in your terminal, run the `data_proc.js` file using the following command:

    ```
    node data_proc.js
    ```

    This outputs the following:

```
exit code.
→ src git:(main) x node data-proc.js
node-pre-gyp info This Node instance does not support builds for N-API version 8
node-pre-gyp info This Node instance does not support builds for N-API version 8
2021-07-19 10:11:02.358971: I tensorflow/core/platform/cpu_feature_guard.cc:142] This TensorFlow binary is optimized with oneAPI Deep Neural Network Library (oneDNN) to use the following CPU i
structions in performance-critical operations:  AVX2 FMA
To enable them in other operations, rebuild TensorFlow with the appropriate compiler flags.
```

|   | Item_Identifier | Item_Weight | Item_Fat_Content | Item_Visibility | ... | Outlet_Size | Outlet_Locati... | Outlet_Type | Item_Outlet_S... |
|---|---|---|---|---|---|---|---|---|---|
| 0 | FDA15 | 9.3 | Low Fat | 0.016047301 | ... | Medium | Tier 1 | Supermarket T... | 3735.138 |
| 1 | DRC01 | 5.92 | Regular | 0.019278216 | ... | Medium | Tier 3 | Supermarket T... | 443.4228 |
| 2 | FDN15 | 17.5 | Low Fat | 0.016760075 | ... | Medium | Tier 1 | Supermarket T... | 2097.27 |
| 3 | FDX07 | 19.2 | Regular | 0 | ... | NaN | Tier 3 | Grocery Store | 732.38 |
| 4 | NCD19 | 8.93 | Low Fat | 0 | ... | High | Tier 3 | Supermarket T... | 994.7052 |

Figure 10.9 – Displaying the head value of the BigMart sales dataset

6.  From the analysis in the Dnotebook file, we noticed that two columns, `Item_Weight` and `Outlet_Sales`, have missing values. In the following code snippet, we'll fill these missing values using the mean and the modal value respectively:

```
...
  salesDf.fillna({
        columns: ["Item_Weight", "Outlet_Size"],
        values: [salesDf['Item_Weight'].mean(),
"Medium"],
        inplace: true
      })
...
```

7.  As we have noticed, the dataset is a mixture of categorical (strings) columns and numeric (`float32` and `int32`) columns. This means we must convert all categorical columns to numeric form before we can pass them to our model. In the following code snippet, we use Danfo.js's `LabelEncoder` to encode each categorical column to a numeric one:

```
...
    let encoder = new dfd.LabelEncoder()
    let catCols = salesDf.select_dtypes(includes =
['string']).column_names // get all categorical column
names
    catCols.forEach(col => {
        encoder.fit(salesDf[col])
        enc_val = encoder.transform(salesDf[col])
        salesDf.addColumn({ column: col, value: enc_val
})
      })
...
```

8.  Next, we'll separate the target from the training dataset. The target, as we have noticed from the problem statement, is the sales price. This corresponds to the last column, `Item_Outlet_Sales`. In the following code snippet, we'll split the dataset using the `iloc` function:

    ```
    . . .
        let Xtrain, ytrain;
        Xtrain = salesDf.iloc({ columns:
    [`1:${salesDf.columns.length - 1}`] })
        ytrain = salesDf['Item_Outlet_Sales']
        console.log(`Training Dataset Shape: ${Xtrain.
    shape}`)
    . . .
    ```

9.  Next, we'll standardize our dataset. Standardizing our dataset forces every column to be in the scale, and as such improves model training. In the following code snippet, we use Danfo.js's `StandardScaler` to standardize the dataset:

    ```
        . . .
    let scaler = new dfd.MinMaxScaler()
        scaler.fit(Xtrain)
        Xtrain = scaler.transform(Xtrain)
    . . .
    ```

10. Finally, to complete the `processData` function, we'll return the raw tensors, as shown in the following code snippet:

    ```
        . . .
        return [Xtrain.tensor, ytrain.tensor]
    . . .
    ```

> **Note**
> You can see the full code in the code repository here: `https://github.com/PacktPublishing/Building-Data-Driven-Applications-with-Danfo.js/blob/main/Chapter10/sales_predictor/src/data-proc.js`.

Executing and printing the tensors from the final `data_proc.js` file should give you tensors like those shown in the following screenshot:

```
Training Dataset Shape: 8523,11
Tensor
    [[0        , 0.2825246, 0    , 0.0488665, 0        , 0.9275072, 0        , 0.5833333, 0  , 0  , 0        ],
     [0.0006418, 0.0812742, 0.25, 0.0587051, 0.0666667, 0.0720684, 0.1111111, 1        , 0  , 0.5, 0.3333333],
     [0.0012837, 0.7707651, 0    , 0.051037 , 0.1333333, 0.4682884, 0        , 0.5833333, 0  , 0  , 0        ],
     ...,
     [0.2272144, 0.3599286, 0    , 0.1071475, 0.6      , 0.2284922, 0.8888889, 0.7916667, 1  , 1  , 0        ],
     [0.5827985, 0.1580828, 0.25, 0.4422188, 0.4      , 0.3049393, 0.1111111, 1        , 0  , 0.5, 0.3333333],
     [0.296534 , 0.610003 , 0    , 0.1366611, 0.0666667, 0.1875098, 0.7777778, 0.5      , 1  , 0  , 0        ]]
Tensor
    [3735.1379395, 443.4227905, 2097.2700195, ..., 1193.1136475, 1845.5976563, 765.6699829]
→ src git:(main) ✗ ▊
```

Figure 10.10 – Final BigMart data tensors after processing

Now you have a function that can process a raw dataset and return tensors, let's move on to creating models with `tfjs`.

# Creating models with TensorFlow.js

As we mentioned earlier, `tfjs` provides a Layers API that can be used to define and create ML models. The Layers API is similar to the popular Keras API, and as such, Python developers already familiar with Keras can easily port their code to `tfjs`.

The Layers API provides two ways of creating models—a sequential and a model format. We'll briefly explain and give examples of these in the following sub-sections.

## The sequential way of creating models

This is the easiest and most common way of creating models. It is simply a stack of multiple model layers, where the first layer in the stack defines the input and the last layer defines the output, while the middle layers can be as many as required.

An example of a two-layer sequential model is shown in the following code snippet:

```
const model = tf.sequential();
// First layer must have an input shape defined.
model.add(tf.layers.dense({units: 32, inputShape: [50]}));
model.add(tf.layers.dense({units: 24}));
model.add(tf.layers.dense({units: 1}));
```

You will notice from the preceding code snippet that the first layer in the sequence provides an `inputShape` parameter. This means that the model is expecting an input with 50 columns.

You can also create a sequential layer by passing a list of layers, as demonstrated in the following code snippet:

```
const model = tf.sequential({
    layers: [tf.layers.dense({units: 32, inputShape: [50]}),
             tf.layers.dense({units: 24}),
             tf.layers.dense({units: 1})]
});
```

Next, let's look at the model format.

## The model way of creating models

The model format of creating models provides more flexibility when creating models. Instead of simply accepting a stack of linear layers, models defined with the model layer can be non-linear, cyclical, and as advanced or connected as you want.

For example, in the following code snippet, we create a two-layer network using the model format:

```
const input = tf.input({ shape: [5] });
const denseLayer1 = tf.layers.dense({ units: 16, activation:
'relu' });
const denseLayer2 = tf.layers.dense({ units: 8, activation:
'relu' });
const denseLayer3 = tf.layers.dense({ units: 1 })
const output = denseLayer3.apply(denseLayer2.apply(denseLayer1.
apply(input)))
const model = tf.model({ inputs: input, outputs: output });
```

From the preceding example code, you can see that we are explicitly calling the `apply` function and passing the layer we want to connect as a parameter. This way, we can build hybrid and highly complex models that have graph-like connections.

You can learn more about the Layers API from the official `tfjs` documentation here: `https://js.tensorflow.org/api/latest/#Models`.

Now you know how to create models with the Layer API, we are going to create a simple three-layer regression model in the next section.

# Creating a simple three-layer regression model

A regression model, as we have explained in the previous chapter, *Chapter 9, Basics of Machine Learning*, is a model with a continuous output. To create a regression model with tfjs, we define stacks of layers, and in the last layer, we set the number of units to 1. For example, open the model.js file in the code repository. On *lines 7-11*, you should see the following sequential model definition:

```
...
const model = tf.sequential();
model.add(tf.layers.dense({ inputShape: [11], units: 128,
kernelInitializer: 'leCunNormal' }));
model.add(tf.layers.dense({units: 64, activation: 'relu' }));
model.add(tf.layers.dense({units: 32, activation: 'relu' }));
model.add(tf.layers.dense({units: 1}))
...
```

Notice that in the first layer, we set the inputShape parameter to 11. This is because we have 11 training columns in our BigMart dataset. You can confirm this by printing the shape of the processed tensors. In the last layer, we set the units property to 1 because we want to predict a single continuous value.

The layers in between can be as many we want, and the units can take on any number. So, in essence, adding more layers in between gives us a deeper model, and adding more units gives us a wider model. The choice of layers to use will depend not only on the problem but also on performing multiple experiments and training.

And with just those few lines of code, you have successfully created a three-layer regression model in tfjs.

After creating a model, the next thing you would normally do is compile the model. So, what does compilation do? Well, compilation is the process of preparing a model for training and evaluation. This means that, in the compilation stage, we have to set the model's optimizers, loss, and/or training metrics.

A tfjs model must be compiled before you can start training. So, how do we compile a model in tfjs? This can be done by calling the compile function on a defined model and setting the optimizer and metrics you want to calculate.

On *lines 13-17* of the `model.js` file, we compiled our regression model by setting the optimizer to `Adam`, and the `loss` and `metrics` properties to `meanSquaredError`. See the following code snippet to view this:

```
...
    model.compile({
        optimizer: tf.train.adam(LEARNING_RATE),
        loss: tf.losses.meanSquaredError,
        metrics: ['mse']
    });
...
```

It is worth mentioning that there are different types of optimizers to pick from; see a full list at `https://js.tensorflow.org/api/latest/#Training-Optimizers`. The choice of which optimizer to use will depend on your experience, as well as on multiple experimentations.

In terms of loss, the problem will inform you on which loss function to use. In our case, since it's a regression problem, we can use the **mean squared error** (**MSE**) function. To see a full list of available loss functions, visit `https://js.tensorflow.org/api/latest/#Training-Losses`.

And finally, in terms of metrics that are calculated and displayed during model training, we can specify multiple options, and just as with the loss, the specified metric will depend on the problem you are trying to solve. In our case, we can also calculate an MSE. To see the full list of supported metrics, visit `https://js.tensorflow.org/api/latest/#Metrics`.

Now you have defined and compiled the model, we will move on to the next and final part of this chapter, which is about training the model.

## Training the model with the processed dataset

The `train.js` file contains the code for training the three-layer regression model on the processed dataset. In the following steps, we'll walk you through the whole process of model training:

1.  First, let's load and process the dataset using the `processData` function, as follows:

    ```
    ...
    const data = await processData("./dataset/train.csv")
    ```

```
const Xtrain = data[0]
const ytrain = data[1]
...
```

2. Next, we load the model using the `getModel` function, as follows:

```
...
const model = getModel()
...
```

3. Next, and very importantly, we call the `fit` function on the model, pass the training data, the target, and a couple of parameters such as the `epoch`, `batchSize`, and `validationSplits` parameters, and a callback function called `onEpochEnd`, as follows:

```
...
    await model.fit(Xtrain, ytrain, {
        batchSize: 24,
        epochs: 20,
        validationSplit: 0.2,
        callbacks: {
            onEpochEnd: async (epoch, logs) => {
                const progressUpdate = `EPOCH (${epoch +
1}): Train MSE: ${Math.sqrt(logs.mse)}, Val MSE:  ${Math.
sqrt(logs.val_mse)}\n`
                console.log(progressUpdate);
            }
        }
    });
...
```

Let's understand what the parameters we passed to the `fit` function do, as follows:

- `Xtrain`: The training data.
- `ytrain`: The target data.
- `epoch`: The epoch size is the number of times to iterate over the training data.
- `batchSize`: The batch size is the number of data points or samples used in computing one gradient update.

- validationSplit: The validation split is a handy parameter that tells tfjs to reserve the specified percentage of data for validation. This can be used when we do not want to manually split our dataset into train and test sets.

- callbacks: A callback, as the name suggests, accepts a list of functions that are called during different life cycles of the model training. Callbacks are important in monitoring model training. See a full list of callbacks here: https://js.tensorflow.org/api/latest/#tf.Sequential.fitDataset.

4. Finally, we save the model so that we can use it in making new predictions, as follows:

```
...
        await model.save("file://./sales_pred_model")
...
```

Running the train.js file will load and process the dataset, load the model, and run the model training for the specified number of epochs. The callback (onEpochEnd) we have specified will print out the loss and the root MSE after each epoch, as shown in the following screenshot:

```
Epoch 14 / 20
eta=0.0 ==================================================>
2129ms 312us/step - loss=1180127.63 mse=1180127.63 val_loss=1240288.13 val_mse=1240288.13
EPOCH (14): Train MSE: 1086.336791699517, Val MSE:  1113.6822369958138

Epoch 15 / 20
eta=0.0 ==================================================>
2040ms 299us/step - loss=1200394.13 mse=1200394.13 val_loss=1224141.75 val_mse=1224141.75
EPOCH (15): Train MSE: 1095.6249928693667, Val MSE:  1106.4093952963342

Epoch 16 / 20
eta=0.0 ==================================================>
2113ms 310us/step - loss=1188992.13 mse=1188992.13 val_loss=1178725.00 val_mse=1178725.00
EPOCH (16): Train MSE: 1090.409154858854, Val MSE:  1085.6910241869

Epoch 17 / 20
eta=0.0 ==================================================>
2040ms 299us/step - loss=1180922.75 mse=1180922.75 val_loss=1206503.00 val_mse=1206503.00
EPOCH (17): Train MSE: 1086.702696232967, Val MSE:  1098.4093044034178

Epoch 18 / 20
eta=0.0 ==================================================>
2074ms 304us/step - loss=1198624.38 mse=1198624.38 val_loss=1184863.50 val_mse=1184863.50
EPOCH (18): Train MSE: 1094.8170509267748, Val MSE:  1088.514354521795

Epoch 19 / 20
eta=0.0 ==================================================>
2060ms 302us/step - loss=1180541.25 mse=1180541.25 val_loss=1213894.63 val_mse=1213894.63
EPOCH (19): Train MSE: 1086.527151064344, Val MSE:  1101.7688618762104

Epoch 20 / 20
eta=0.0 ==================================================>
2023ms 297us/step - loss=1179258.13 mse=1179258.13 val_loss=1159443.75 val_mse=1159443.75
EPOCH (20): Train MSE: 1085.9365197837303, Val MSE:  1076.7746978825237
```

Figure 10.11 – Model training logs showing loss and root MSE

And that's it! You have successfully created, trained, and saved a regression model that can predict sale prices using TensorFlow.js. In the next and final section of this chapter, we'll show you how to load your saved model and use it to make predictions.

# Making predictions with the trained model

In order to make predictions, we have to load the saved model and call the `predict` function on it. TensorFlow.js provides a `loadLayersModel` function that can be used to load saved models from a filesystem. In the following steps, we show you how to achieve this:

1.  Create a new file called `predict.js`.

2.  In the `predict.js` file, add the following code:

```
const dfd = require("danfojs-node")
const tf = dfd.tf
async function loadModel() {
    const model = await tf.loadLayersModel('file://./
sales_pred_model/model.json');
    model.summary()
    return model
}
loadModel()
```

The preceding code loads the saved model from the file path and prints a summary. The output of the summary should be similar to that shown in the following screenshot:

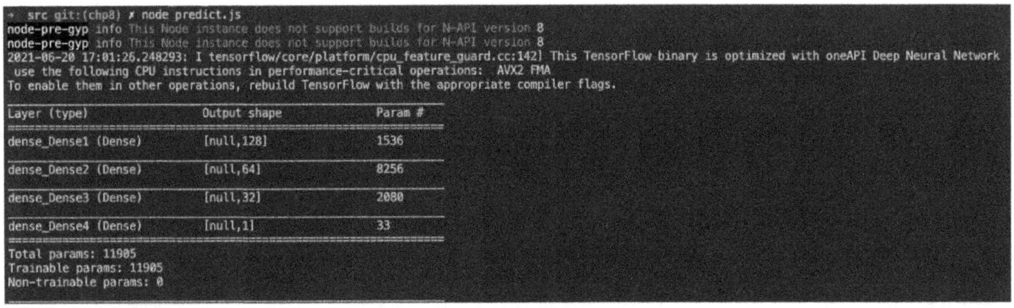

Figure 10.12 – Model summary of the saved model

3. Now, create a new function called `predict` that uses the saved model to make a prediction, as shown in the following code snippet:

```
. . .
async function predict() {
    //You'll probably have to do some data pre-processing
as we did before training
    const data = [0.1, 0.21, 0.25, 0.058, 0.0, 0.0720,
0.111, 1, 0, 0.5, 0.33] //sample processed test data
    const model = await loadModel()
    const value = model.predict(tf.tensor(data, [1, 11]))
//cast data to required shape
    console.log(value.arraySync());


}
predict()
```

The output is shown here:

```
[ [ 738.65380859375 ] ]
. . .
```

In the preceding function, we call the `predict` function on the model and pass a tensor with the correct shape (batch, 11) our model is expecting. This returns a tensor of the prediction and from this tensor, we get the underlying value. From this, we can tell a product with those specific values will sell for approximately **US Dollars (USD)** $739.

> **Note**
>
> In a real-world application, you would generally load a test dataset from another **comma-separated values (CSV)** file and apply the same data processing steps as we did during training. The example uses an inline data point, just to demonstrate using a saved model to make predictions.

That brings us to the end of this chapter! Congratulations on making it this far. I'm sure you have learned a lot. In the next chapter, we'll go deeper by building a more practical application—a recommendation system!

# Summary

In this chapter, we introduced you to the basics of TensorFlow.js. Specifically, you learned how to install TensorFlow.js both in the browser and Node.js environment, you learned about tensors and the core data structure of `tfjs`, you learned about the Core and Layer APIs, and finally, you learned how to build, train, and save a regression model.

In the next chapter, we'll go deeper into a more practical and hands-on project, and the knowledge gained here will help you build great products with TensorFlow.js and Danfo.js.

# 11
# Building a Recommendation System with Danfo.js and TensorFlow.js

In the preceding chapter, we introduced you to TensorFlow.js and showed you how to create a simple regression model to predict sales prices. In this chapter, we'll take this a step further by creating a recommendation system that can recommend movies for different users while taking into account a user preference. By the end of this chapter, you will understand how recommendation systems work, as well as how to build one with JavaScript.

Specifically, we'll cover the following topics:

- What is a recommendation system?
- The neural network approach to creating a recommendation system
- Building a movie recommendation system

# Technical requirements

To follow along in this chapter, you will need the following:

- A modern browser such as Chrome, Safari, Opera, or Firefox

- **Node.js**, **Danfo.js**, TensorFlow.js, and (optionally) **Dnotebook** installed on your system

- A stable internet connection for downloading a dataset

- The code for this chapter is available and can be cloned from GitHub at `https://github.com/PacktPublishing/Building-Data-Driven-Applications-with-Danfo.js/tree/main/Chapter11`

The installation instructions for Danfo.js, TensorFlow.js, and Dnotebook can be found in *Chapter 3, Getting Started with Danfo.js, Chapter 10, Introduction to TensorFlow.js*, and *Chapter 2, Dnotebook – An Interactive Computing Environment for JavaScript*, respectively.

# What is a recommendation system?

A **recommendation system** is any system that can predict a preference or a usefulness score given to an item by a user. Using this preference score, it can recommend items to the user.

Items here can be digital products such as movies, music, books, and even clothes. The goal of every recommendation system is to be able to recommend items that the user will like.

Recommendation systems are very popular and can be found almost everywhere; for example:

- Movie streaming platforms such as *Netflix, Amazon Prime, Hulu*, and *Disney+* use recommendation systems for recommending movies to you.

- Social media websites such as *Facebook, Twitter*, and *Instagram* use recommendation systems for recommending friends to connect with.

- E-commerce websites such as *Amazon* and *AliExpress* use recommendation systems for recommending products such as clothes, books, and electronics.

Recommendation systems are mostly built using data from user-item interaction. Hence, there are three major approaches that are followed when building recommendation systems. These are **collaborative filtering**, **content-based filtering**, and the **hybrid approach**. We'll briefly explain these approaches in the following sub-sections.

# Collaborative filtering approach

In the collaborative filtering approach, the recommendation system is modeled based on users' past behavior or history. That is, this approach leverages existing user interactions such as ratings, likes, or reviews for items to model the user's preference, thereby understanding what the user likes. The following diagram shows how the collaborative filtering approach helps with building a recommendation system:

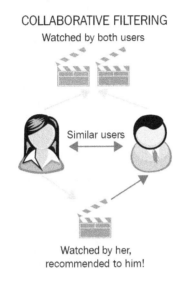

Figure 11.1 – Collaborative filtering approach to building a recommendation system

In the preceding diagram, you can see that two users who have seen the same movie, and probably rated the same way, are grouped as similar users, since the movie that's been seen by the person on the left has been recommended to the person on the right. In the content-based filtering approach, the recommendation system is modeled based on **item characteristics**. That is, items may be pre-tagged with certain characteristics such as category, price, genre, size, and ratings received, and using these characteristics, the recommendation system can recommend similar items.

The following diagram shows how the content-based filtering approach to building a recommendation system works:

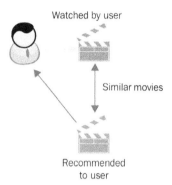

Figure 11.2 – Content-based filtering approach to building a recommendation system

In the preceding diagram, you can observe that movies similar to each other get recommended to a user.

## Hybrid filtering approach

The hybrid approach, as its name suggests, is a combination of the collaborative and content-based filtering approaches. That is, it combines the best of both approaches to create an even better recommendation system. The majority of real-world recommendation systems today use this approach to mitigate the shortcomings of individual approaches.

The following diagram shows one way of combining the content-based filtering approach with the collaborative filtering approach to create a hybrid recommendation system:

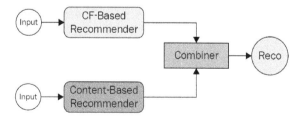

Figure 11.3 – The hybrid approach to building a recommendation system

In the preceding diagram, you can see that we have two inputs feeding the hybrid system. These inputs go into the collaborative (**CF**) and content-based systems, and then the output from these systems gets combined. This combination can be customized and can even serve as input to other advanced systems such as neural networks. The overall goal here is to create a powerful hybrid system by combining multiple recommendation systems.

It is worth mentioning that any approach that's used in creating a recommendation system will need one form of data or the other. For example, in the collaborative filtering approach, you will need the *user-item interaction* history, while for the content-based approach, you'll need *item metadata*.

If you have enough data to train a recommendation system, you can leverage numerous machine and non-machine learning techniques to model the data before making recommendations. Some popular algorithms you can use are **K-nearest neighbor** (`https://en.wikipedia.org/wiki/K-nearest_neighbors_algorithm`), **clustering algorithms** (`https://en.wikipedia.org/wiki/Cluster_analysis`), **decision trees** (`https://en.wikipedia.org/wiki/Decision_trees`), **Bayesian classifiers** (`https://en.wikipedia.org/wiki/Naive_Bayes_classifier`), and even **artificial neural networks** (`https://en.wikipedia.org/wiki/Artificial_neural_networks`).

In this chapter, we'll use the **neural network** approach to build a recommendation system. We'll explain this in detail in the following section.

# The neural network approach to creating a recommendation system

In recent years, neural networks have been the Swiss army knife when it comes to solving many problems in the field of **machine learning** (**ML**). This is evident in areas of ML breakthroughs such as **image classification/segmentation** and **natural language processing**. With the availability of data, neural networks have been successfully used to build large-scale recommendation systems such as the ones used at *Netflix* (`https://research.netflix.com/research-area/machine-learning`) and *YouTube* (`https://research.google/pubs/pub45530/`).

Although there are different approaches to building a recommendation system with neural networks, they all rely on one major fact: they need an efficient way to learn similarities between items or users. In this chapter, we'll leverage a concept called **embeddings** to efficiently learn these similarities so that we can easily power our recommendation system.

But first, what are embeddings and why are we using them? In the next sub-section, we'll briefly answer these questions.

## What is an embedding?

An embedding is a mapping of discrete variables to continuous or real-valued variables. That is, given a set of variables such as [good, bad], an embedding can map each discrete item to a continuous vector of *n* dimensions – for example, good can be represented as [0.1, 0.6, 0.1, 0.8] and bad can be represented as [0.8, 0.2, 0.6, 0.1], as shown in the following figure:

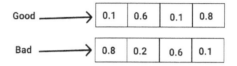

Figure 11.4 – Representing discrete categories with real-valued variables

If you're familiar with encoding schemes such as **one-hot encoding** (https://en.wikipedia.org/wiki/One-hot) or **label encoding** (https://machinelearningmastery.com/one-hot-encoding-for-categorical-data/), then you might be wondering how embedding is different from them.

Well, there are two major differences and, technically, advantages of embedding:

- Embedding representations can be small or large, depending on the specified dimension. This is different from encoding schemes such as one-hot encoding, where the dimensionality of the representation increases with the number of discrete classes.

  For example, the following figure shows how the dimension that's used by the one-hot encoding representation increases with the number of unique countries:

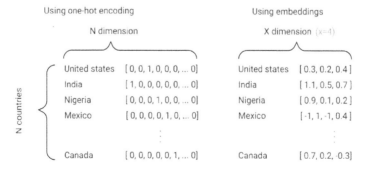

Figure 11.5 – Size comparison between embeddings and one-hot encoding

- Embeddings can be learned alongside weights in neural networks. This is the major advantage over other encoding schemes because, with this property, learned embeddings become a similarity cluster of discrete classes, which means that you can easily find similar items or users. For example, looking at the following proofed, you can see that we have two groups of learned word embeddings:

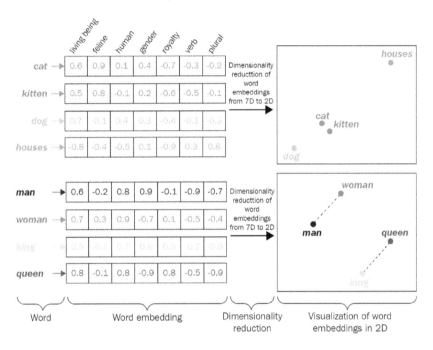

Figure 11.6 – Embedding words and showing similarity in an embedding space (Redrawn from: https://medium.com/@hari4om/word-embedding-d816f643140)

In the preceding figure, you can see that the group depicting **man**, **woman**, **king**, and **queen** is being passed through an embedding and that the resulting output is an embedding space, where words that are closer in meaning are grouped. This is made possible by the learned word *embedding*.

So, how do we leverage embeddings to create a recommendation system? Well, as we mentioned previously, embeddings can efficiently represent data, which means we can use them to learn or represent user-item interaction. As such, we can easily use learned embeddings to find similar items to recommend. We can even take this further by combining an embedding with a supervised machine learning task.

This approach of combining a learned embedding representation with a supervised ML task is what we'll do in the next section to create our movie recommendation system.

# Building a movie recommendation system

To build a movie recommendation system, we need some kind of user-movie interaction dataset. Fortunately, we can use the MovieLens 100k dataset (https://grouplens.org/datasets/movielens/100k/) provided by **Grouplens** (https://grouplens.org/). This data contains 100,000 movie ratings given by 1,000 users on 1,700 movies.

The following screenshot shows the first few rows of the dataset:

```
user_id,item_id,rating,timestamp
196,242,3,881250949
186,302,3,891717742
22,377,1,878887116
244,51,2,880606923
166,346,1,886397596
298,474,4,884182806
115,265,2,881171488
253,465,5,891628467
305,451,3,886324817
6,86,3,883603013
62,257,2,879372434
286,1014,5,879781125
200,222,5,876042340
210,40,3,891035994
224,29,3,888104457
303,785,3,879485318
122,387,5,879270459
194,274,2,879539794
291,1042,4,874834944
234,1184,2,892079237
119,392,4,886176814
167,486,4,892738452
```

Figure 11.7 – The first few rows of the MovieLens dataset

From the preceding screenshot, you can see that we have user_id, item_id (movies), and the rating the user gave to the item (movie). With just this interaction and the use of embeddings, we can efficiently model the behavior of users and, as such, learn what kind of movies they like.

To understand how we'll build and learn this interaction with embeddings and a neural network, please refer to the following architecture diagram:

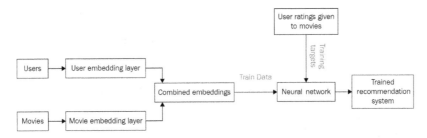

Figure 11.8 – High-level architecture of our recommendation system

From the preceding diagram, you can see that we have two embedding layers, one for the users and the other for the items (movies). These two embedding layers are then combined before being passed to a dense layer.

So, in essence, we are combining embeddings with a supervised learning task, where the output from the embeddings gets passed to a dense layer to predict ratings a user will give an item (movie).

You may be wondering, if we are learning to predict ratings a user will give a product, how does this help us make recommendations? Well, the trick is that if we can efficiently predict the rating a user will give a movie, then, using the learned similarity embedding, we can predict the rating the user will give to all movies. Then, with this information, we can recommend the movies with the highest predicted ratings to the user.

So, how can we build this seemingly complicated recommendation system in JavaScript? Well, in the next sub-section, we'll show you how you can easily use TensorFlow.js, combined with Danfo.js, to achieve this.

## Setting up your project directory

The code and the dataset you need to successfully follow this chapter are provided in the code repository for this chapter (https://github.com/PacktPublishing/Building-Data-Driven-Applications-with-Danfo.js/tree/main/Chapter11).

You can download the entire project to your computer to easily follow along. If you have downloaded the project code, then navigate to your root directory, where the src folder is visible.

Inside the src folder, you have the following folders/scripts:

- book_recommendation_model: This is the folder where the trained model is saved to.
- data: This folder contains our training data.

- `data_proc.js`: This script contains all our data processing code.

- `model.js`: This script defines and compiles the recommendation model.

- `recommend.js`: This script contains code for making recommendations.

- `train.js`: This script contains code for training the recommendation model.

To quickly test the pre-trained recommendation model, first, install all the necessary packages using `yarn` (recommended) or NPM, and then run the following command:

```
yarn recommend
```

This will recommend 10, 5, and 20 movies for users with user IDs of 196, 880, and 13. If successful, you should see an output similar to the following:

```
[
  'Remains of the Day, The (1993)',
  'Star Trek: First Contact (1996)',
  'Kolya (1996)',
  'Men in Black (1997)',
  'Hunt for Red October, The (1990)',
  'Sabrina (1995)',
  'L.A. Confidential (1997)',
  'Jackie Brown (1997)',
  'Grease (1978)',
  'Dr. Strangelove or: How I Learned to Stop Worrying and Love the Bomb (1963)'
]
[
  'Legends of the Fall (1994)',
  'Kolya (1996)',
  'L.A. Confidential (1997)',
  'Jackie Brown (1997)',
  'Dr. Strangelove or: How I Learned to Stop Worrying and Love the Bomb (1963)'
]
[
  'Batman Forever (1995)',
  'To Wong Foo, Thanks for Everything! Julie Newmar (1995)',
  'Legends of the Fall (1994)',
  'Remains of the Day, The (1993)',
  'Star Trek: First Contact (1996)',
  'Kolya (1996)',
  'Men in Black (1997)',
  'Hunt for Red October, The (1990)',
  'Sabrina (1995)',
  'L.A. Confidential (1997)',
  'Jackie Brown (1997)',
  'Age of Innocence, The (1993)',
  'Man Without a Face, The (1993)',
  'Grease (1978)',
  'Jungle Book, The (1994)',
  'Dr. Strangelove or: How I Learned to Stop Worrying and Love the Bomb (1963)',
  'Only You (1994)',
  "Romy and Michele's High School Reunion (1997)",
  'Just Cause (1995)',
  'Endless Summer 2, The (1994)'
]
✨  Done in 7.91s.
```

Figure 11.9 – Recommended movies provided by the trained recommendation system

You can also retrain the model by running the following command:

```
yarn retrain
```

The preceding command, by default, will retrain the model with a batch size of 128, an epoch size of 5, and, when done, will save the trained model to the book_ recommender_model folder.

Now that you have set up the project locally, we'll work through each section and explain how to build the recommendation system from scratch.

## Retrieving and processing the training dataset

The dataset we're using was retrieved from the Grouplens website. By default, the movies data (https://files.grouplens.org/datasets/movielens/ml-100k.zip) is a ZIP file containing tab-separated files. For the sake of simplicity, I have downloaded and converted  the two files you'll need for this project into CSV format. You can get these files from the data folder in this project's code repository (https://github.com/ PacktPublishing/Building-Data-Driven-Applications-with-Danfo. js/tree/main/Chapter11/src/data).

There are two main files:

- movieinfo.csv: This file contains metadata about each movie, such as its title, description, and links.

- movielens.csv: This is the user ratings dataset.

To use the movielens dataset, we must read the dataset with Danfo.js, process it, and then convert it into a tensor that we can pass to our neural network.

In your project's code, open the data_proc.js script. This script exports one main function called processData with the following code:

```
. . .
    const nItem = (moviesDF["item_id"]).max()
    const nUser = (moviesDF["user_id"]).max()

    const moviesIdTrainTensor = (moviesDF["item_id"]).tensor
    const userIdTrainTensor = (moviesDF["user_id"]).tensor
    const targetData = (moviesDF["rating"]).tensor

    return {
```

```
            trainingData: [moviesIdTrainTensor, userIdTrainTensor],
            targetData,
            nItem,
            nUser
    }
    ...
```

So, what are we doing in the preceding code? Well, thankfully, we do not need much data pre-processing because the user_id, item_id, and ratings columns are already in numeric form. So, we are simply doing two things:

- Retrieving the maximum IDs from the item and user columns. This number, called the **vocabulary size**, will be passed to the embedding layer when creating our model.

- Retrieving and returning the underlying tensors of the user, item, and rating columns. The user and item tensors will serve as our training input, while the rating tensors will become our supervised learning target.

Now that you know how to process the data, let's start building the neural network using TensorFlow.js.

## Building the recommendation model

The complete code for our recommendation model can be found in the model.js file. This model uses a hybrid approach, as we saw in the high-level architecture diagram (see *Figure 11.8*).

> **Note**
>
> We are using the Model API we introduced in *Chapter 10, Introduction to TensorFlow.js*, to create the network. This is because we are creating a complex architecture and we need more control in terms of inputs and outputs.

In the following steps, we will explain the model and show the corresponding code for creating it:

1. **The inputs**: Every model will have an entry point, and in our case, we have two inputs: one from user and the other from the items:

```
    ...
    const itemInput = tf.layers.input({ name: "itemInput",
    shape: [1] })
```

```
const userInput = tf.layers.input({ name: "userInput",
shape: [1] })
. . .
```

Notice that the input has a shape parameter set to 1. This is because our input tensors are vectors with a dimension of 1.

2.  **The embedding layers**: Now, we need to create two separate embedding layers – one for the user and the other for the movies:

```
. . .
const itemEmbedding = tf.layers.embedding({
        inputDim: nItem + 1,
        outputDim: 16,
        name: "itemEmbedding",
    }).apply(itemInput)

const userEmbedding = tf.layers.embedding({
        inputDim: nUser + 1,
        outputDim: 16,
        name: "userEmbedding",
    }).apply(userInput)
. . .
```

The embedding layers accept the following parameters:

a) `InputDim`: This is the vocabulary size of the embedding vector. The maximum integer index is + 1.

b) `OutputDim`: This is a user-specified output dimension. That is, it is used to configure the size of the embedding vector.

Next, we'll merge these embedding layers.

3.  **Merging the embedding layers**: In the following code block, we are merging the output from the user and item embedding layer using a `dot` product, flattening the output, and passing the output to a dense layer:

```
. . .
const mergedOutput = tf.layers.dot({ axes: 0}).
apply([itemEmbedding, userEmbedding])
const flatten = tf.layers.flatten().apply(mergedOutput)
const denseOut = tf.layers.dense({ units: 1,
```

```
activation: "sigmoid", kernelInitializer: "leCunUniform"
}).apply(flatten)
...
```

With the preceding output, we can now define our model using the Models API.

4.  Finally, we will define and compile the model, as shown in the following code snippet:

```
...
const model = tf.model({ inputs: [itemInput, userInput],
outputs: denseOut })
        model.compile({
            optimizer: tf.train.adam(LEARNING_RATE),
            loss: tf.losses.meanSquaredError
        });
...
```

The preceding code uses the Models API to define the inputs and output, and then calls the compile method, which accepts a training optimizer (Adam optimizer) and a loss function (mean squared error). You can view the full model code in the model.js file.

With the model architecture defined, we can start training the model.

## Training and saving the recommendation model

The training code for the model can be found in the train.js file. This code has two main sections. We'll look at both here.

The first section, shown in the following code block, trains the model using a batch size of 128, an epoch size of 5, and a validation split of 0.1-10% for the training data, which is reserved for model validation:

```
...
await model.fit(trainingData, targetData, {
        batchSize: 128,
        epochs: 5,
        validationSplit: 0.1,
        callbacks: {
```

```
        onEpochEnd: async (epoch, logs) => {
            const progressUpdate = `EPOCH (${epoch + 1}):
Train MSE: ${Math.sqrt(logs.loss)}, Val MSE:  ${Math.sqrt(logs.
val_loss)}\n`
            console.log(progressUpdate);
        }
    }
  });
...
```

In the preceding training code, we printed the loss after each training epoch. This helps us track training progress.

The following code block saves the trained model to the file path provided. In our case, we are saving it to the `movie_recommendation_model` folder:

```
...
await model.save(`file://${path.join(__dirname, "movie_
recommendation_model")}`)
...
```

Take note of this folder's name as we'll be using it when we make recommendations in the next sub-section.

To train the model, you can run the following command in the `src` folder:

```
yarn train
```

Alternatively, you can directly run `train.js` with `node`:

```
node train.js
```

This will start model training for the specified number of epochs and, once done, save the model to the specified folder. Once the training is complete, you should have an output similar to the following:

```
Training started....
Epoch 1 / 5
eta=0.0 ===========================================================>
5154ms 57us/step - loss=7.95 val_loss=7.67
EPOCH (1): Train MSE: 2.8189935109583772, Val MSE:  2.7695004309485

Epoch 2 / 5
eta=0.0 ===========================================================>
4697ms 52us/step - loss=7.67 val_loss=7.67
EPOCH (2): Train MSE: 2.769029149434856, Val MSE:  2.7694734855191174

Epoch 3 / 5
eta=0.0 ===========================================================>
4867ms 54us/step - loss=7.67 val_loss=7.67
EPOCH (3): Train MSE: 2.768992986414642, Val MSE:  2.7694692672013272

Epoch 4 / 5
eta=0.0 ===========================================================>
4547ms 51us/step - loss=7.67 val_loss=7.67
EPOCH (4): Train MSE: 2.768991522663402, Val MSE:  2.769468406319356

Epoch 5 / 5
eta=0.0 ===========================================================>
4573ms 51us/step - loss=7.67 val_loss=7.67
EPOCH (5): Train MSE: 2.7689910921481844, Val MSE:  2.769468320231144

Saving model...
+  Done in 30.13s.
```

Figure 11.10 – Training logs of the recommendation model

Once you have a trained and saved model, you can start making movie recommendations.

# Making movie recommendations with the saved model

The `recommend.js` file contains the code for making recommendations. We have also included a utility function called `getMovieDetails`. This function maps a movie ID to the movie metadata so that we can display useful information, such as the name of a movie.

But how do we make recommendations? Well, since we have trained our model to predict the ratings a user will give a set of movies, we can simply pass a user ID and all the movies to the model to make ratings predictions.

With the ratings predictions for all movies, we can simply sort them in descending order and then return the top movies as recommendations.

To do this, follow these steps:

1.  First, we must get all the unique movie IDs to predict for:

    ```
    ...
    const moviesDF = await dfd.read_csv(moviesDataPath)
    const uniqueMoviesId = moviesDF["item_id"].unique().
    values
    const uniqueMoviesIdTensor = tf.tensor(uniqueMoviesId)
    ...
    ```

2.  Next, we must construct a user tensor that's the same length as the movie ID tensor. The tensor will have the same user ID across all entries because for each movie, we are predicting the rating the same user will give:

    ```
    ...
    const userToRecommendForTensor =
    tf.fill([uniqueMoviesIdTensor.shape[0]], userId)
    ...
    ```

3.  Next, we must load the model and call the `predict` function by passing the movie and user tensors as input:

    ```
    ...
    const model = await loadModel()
        const ratings = model.predict([uniqueMoviesIdTensor,
      userToRecommendForTensor])
    ...
    ```

    This will return a tensor of predicted ratings the user will give to each movie.

4.  Next, we must construct a DataFrame with two columns called `movie_id` (unique movie IDs) and `ratings` (predicted ratings given by the user to each movie):

    ```
    ...
    const recommendationDf = new dfd.DataFrame({
        item_id: uniqueMoviesId,
        ratings: ratings.arraySync()
      })
    ...
    ```

5.  Storing the predicted ratings and the corresponding movie ID in a DataFrame helps us easily sort the ratings, as shown in the following code:

```
. . .
    const topRecommendationsDF = recommendationDf
        .sort_values({
            by: "ratings",
            ascending: false
        })
        .head(top) //return only the top rows
. . .
```

6.  Finally, we must pass the array of sorted movie IDs to the `getMovieDetails` utility function. This function will map each movie ID to the corresponding metadata and return a DataFrame with two columns (movie title and movie release date), as shown in the following code:

```
. . .
    const movieDetailsDF = await
    getMovieDetails(topRecommendationsDF["movie_id"].values)
. . .
```

The `recommend.js` script in the `src` folder contains the full code for making recommendations, including the utility function for mapping a movie ID to its metadata.

To test a recommendation, you need to call the `recommend` function and pass a movie ID and the number of recommendations you want, as shown in the following example:

```
recommend(196, 10) // Recommend 10 movies for user with id 196
```

The preceding code gives us the following output in the console:

```
[
    'Remains of the Day, The (1993)',
    'Star Trek: First Contact (1996)',
    'Kolya (1996)',
    'Men in Black (1997)',
    'Hunt for Red October, The (1990)',
    'Sabrina (1995)',
    'L.A. Confidential (1997)',
    'Jackie Brown (1997)',
```

```
   'Grease (1978)',
   'Dr. Strangelove or: How I Learned to Stop Worrying and Love
the Bomb (1963)'
]
```

And that's it! You have successfully created a recommendation system using neural network embeddings, which can efficiently recommend movies to different users. Using the concepts you've learned about in this chapter, you can easily create different recommendation systems that can recommend different products, such as music, books, and videos.

# Summary

In this chapter, we successfully built a recommendation system that can recommend movies to users based on their preferences. First, we defined what a recommendation model is before briefly talking about the three approaches to designing a recommendation system. Then, we talked about neural network embeddings and why we decided to use them to create our recommendation model. Finally, we put together all the concepts we've learned about by building a movie recommendation model that can recommend the specified number of movies to a user.

With the knowledge you've gained in this chapter, you can easily create a recommendation system that can be embedded in your JavaScript applications.

In the next and final chapter, you'll build another hands-on application using Danfo.js and the **Twitter API**.

# 12
# Building a Twitter Analysis Dashboard

The main goal of this chapter is to show how you can build a full stack web analytics platform using Danfo.js at the backend and the frontend.

To demonstrate this, we will be building a small single-page web app in which you can run a search on a Twitter user, obtain all the tweets in which they are mentioned on a specific day, and perform some simple analysis such as sentiment analysis, drawing insights from the data.

In this chapter, we'll look at the following topics for building the web app:

- Setting up the project environment
- Building the backend
- Building the frontend

# Technical requirements

The following is required for this chapter:

- Knowledge of React.js

- The code for this chapter, which is available here: `https://github.com/PacktPublishing/Building-Data-Driven-Applications-with-Danfo.js/tree/main/Chapter12`

# Setting up the project environment

For this project, we will build a single web page with both a backend and a frontend. We will be using the Next.js framework to build the app. Next.js makes it possible for you to build the backend and frontend quickly and easily. We will also make use of `tailwindcss,` as we have done for some of our previous projects, such as the no-code environment project.

To set up our project environment with Next.js containing the default `tailwindcss` configuration, all we need to do is run the following command:

```
$ npx create-next-app -e with-tailwindcss twitterdashboard
```

The `npx` command runs `create-next-app`, which creates Next.js boilerplate code, including the `tailwindcss` configuration in the `twitterdashboard` directory. Note that the `twitterdashboard` directory (also called *project name*) can be given any name of your choice. If everything is successfully installed, you should get the output shown in the following screenshot:

```
npm run dev
    Starts the development server.

npm run build
    Builds the app for production.

npm start
    Runs the built app in production mode.

We suggest that you begin by typing:

cd twitterdashboard
npm run dev
```

Figure 12.1 – Code environment setup

Now that we are done with the installation, if everything works correctly, you should have the following files in your project:

Figure 12.2 – Directory structure

Finally, to test whether the project is well installed and ready to go, let's run the following command:

```
$ npm run dev
```

This command should automatically start the app and open up the browser, showing the following interface:

Figure 12.3 – Next.js UI

For this project, we will modify the interface shown in *Figure 12.3* to suit our tastes.

Now that the code environment is set up, let's move on to creating our app.

# Building the backend

In this section, we will be looking at how to create the following APIs for our app:

- `/api/tweet`: This API is responsible for fetching a Twitter user and obtaining their data.

- `/api/nlp`: This API is responsible for running sentiment analysis on the obtained user data.

These APIs will be consumed by the frontend components and will be used to create different visualizations and analyses. Let's start by creating the API to fetch a Twitter user's data.

## Building the Twitter API

In this section, we will build an API that makes it easy to obtain tweets in which a Twitter user is mentioned. From each of the tweets, we will obtain their metadata, such as the text, the name of the sender, the numbers of likes and retweets, the device used to tweet, and the time the tweet was created.

To build the Twitter API for fetching a Twitter user's data and structure it to our taste for easy consumption in the frontend, we need to install a tool that makes it easier to interact with the main Twitter developer API. In the following command, we will install `twit.js` for easy access and handling of the Twitter API:

```
$ npm i twit
```

Once `twit` is installed, we need to configure it. To use `twit`, we will need various Twitter developer keys, such as the following:

```
consumer_key='....',
consumer_secret='....',
access_token='.....',
access_token_secret='.....'
```

If you don't have these keys, you will need to create a Twitter developer account and then apply to get access to the API via `https://developer.twitter.com/`. If given access to use the Twitter API, you can visit `https://developer.twitter.com/en/apps` to create an app and set up your credential keys.

> **Note**
>
> Obtaining the Twitter API might take several days, based on how you describe your use case. For a proper step-by-step guide with visual aids to setting up the Twitter API and obtaining the necessary keys, follow the steps here: `https://realpython.com/twitter-bot-python-tweepy/#creating-twitter-api-authentication-credentials`.

After obtaining the Twitter developer keys needed for the project, we'll make use of them in our code. To prevent exposing the keys to the public, let's create a file named `.env.local`, and in this file, let's add our API keys as shown in the following code block:

```
CONSUMER_KEY='Put in your CONSUMER_KEY',
CONSUMER_SECRET='Your CONSUMER_SECRET',
ACCESS_TOKEN ='Your ACCESS_TOKEN',
ACCESS_TOKEN_SECRET='Your ACCESS_TOKEN_SECRET'
```

In Next.js, all APIs are created in a `/pages/api` folder. Next.js uses a file structure in the `pages/` folder to create the URL route. For example, if you have a file named `login.js` located in the `pages/` folder, the content in `login.js` will be rendered in `http://localhost:3000/login`.

The preceding paragraph shows how a route is created for web pages in Next.js based on the filename and structure. The same thing applies to creating APIs in Next.js.

Let's assume we created an API for signing up in `signup.js` located in `pages/api`. This API will automatically be available in `http://localhost:3000/api/signup` and if we were to use this API within the app itself, we could make a call to it like this: `/api/signup`.

For our `/api/tweet` API, let's create a file named `tweet.js` in `pages/api/` and update the file with the following steps:

1. First, we import `twit.js`, and then create a function to clean each of the tweets:

   ```
   const Twit = require('twit')
   ```

```
function clean_tweet(tweet) {
  tweet = tweet.normalize("NFD") //normalize text
  tweet = tweet.replace(/(RT\s(@\w+))/g, '') //remove Rt
tag followed by an @ tag
  tweet = tweet.replace(/(@[A-Za-z0-9]+)(\S+)/g, '') //
remove user name e.g @name
  tweet = tweet.replace(/((http|https):(\S+))/g, '') //
remove url
  tweet = tweet.replace(/[!#?:*%$]/g, '') //remove # tags
  tweet = tweet.replace(/[^\s\w+]/g, '') //remove
punctuations
  tweet = tweet.replace(/[\n]/g, '') //remove newline
  tweet = tweet.toLowerCase().trim() //trim text
  return tweet
}
```

The clean_tweet function takes in tweet text, normalizes the text, removes the hashtag character, users' names, URL links, and newlines, and then trims the text.

2.  We then create a function called twitterApi, which will be used to create our API:

```
export default function twitterAPI(req, res) {
  // api code here
}
```

The twitterApi function takes in two arguments, req and res, which are the server request and response arguments.

3.  We will now update twitterApi with the necessary code:

```
if (req.method === "POST") {
    const { username } = req.body

    const T = new Twit({
        consumer_key: process.env.CONSUMER_KEY,
        consumer_secret: process.env.CONSUMER_SECRET,
        access_token: process.env.ACCESS_TOKEN,
        access_token_secret: process.env.ACCESS_TOKEN_
SECRET,
        timeout_ms: 60 * 1000,   // optional HTTP request
```

```
timeout to apply to all requests.
      strictSSL: true,       // optional - requires SSL
  certificates to be valid.
    })
}
```

First, we check whether the `req.method` request method is a `POST` method, and then we obtain the username from the request body sent via the search box.

The `Twit` class is instantiated and our Twitter API keys are passed in. Since our Twitter developer API keys are stored as environmental keys in `.env.local`, we can easily access each of the keys using `process.env`.

4.  We've configured `twit.js` with our API keys. Let's now do a search of all the tweets that mention a user:

```
T.get('search/tweets', { q: `@${username}`, tweet_mode:
'extended' }, function (err, data, response) {
  let dfData = {
    text: data.statuses.map(tweet => clean_tweet(tweet.
full_text)),
    length: data.statuses.map(tweet => clean_tweet(tweet.
full_text).split(" ").length),
    date: data.statuses.map(tweet => tweet.created_at),
    source: data.statuses.map(tweet => tweet.source.
replace(/<(?:.|\n)*?>/gm, '')),
    likes: data.statuses.map(tweet => tweet.favorite_
count),
    retweet: data.statuses.map(tweet => tweet.retweet_
count),
    users: data.statuses.map(tweet => tweet.user.screen_
name)
  }
  res.status(200).json(dfData)
})
```

We search for all tweets using the `search/tweets` API in the `T.get` method. We then pass in the `param` object containing the username of the user we want to search for.

A dfData object is created to structure the data based on how we want the API output response to be. dfData contains the following keys, which are the metadata we want to extract from the tweet:

- text: The texts in the tweet
- length: The length of the tweet
- date: The date the tweet was tweeted
- source: The device used to create the tweet
- likes: The number of likes the tweet has
- retweet: The number of retweets the tweet has
- users: The user who created the tweet

The metadata in the preceding list is extracted from the JSON data returned from search/tweets in the T.get() method from the preceding code. All the metadata extracted from this JSON data is contained in an array of objects called statuses and the following shows the structure of the JSON data:

```
{
  statuses:[{
     ......
  },
     ......
  ]
}
```

The Twitter API is created and ready to use. Let's move ahead to create the sentiment analysis API.

## Building the text sentiment API

From the /api/tweet API, we are going to obtain the structured JSON data and then perform sentiment analysis.

The sentiment analysis on the data will be fetched via the /api/nlp route. Hence, in this section, we will see how to create a sentiment analysis API for our Twitter data.

Let's create a file named `nlp.js` in the `/pages/api/` folder and update it with the following steps:

1.  We will make use of the `nlp-node.js` package for our sentiment analysis. We'll also make use of `danfojs-node` for data preprocessing, so let's install these two packages:

    ```
    $ npm i node-nlp danfojs-node
    ```

2.  We import `SentimentAnalyzer` from `nlp-node` and `DataFrame` from `danfojs-node`:

    ```
    const { SentimentAnalyzer } = require('node-nlp')
    const { DataFrame } = require("danfojs-node")
    ```

3.  Next, we will create a default `SentimentApi` export function, which will contain our API code:

    ```
    export default async function SentimentApi(req, res) {


    }
    ```

4.  We will then check whether the request method is a `POST` request, and then perform some data preprocessing on the data obtained from the request body:

    ```
    if (req.method === "POST") {
        const sentiment = new SentimentAnalyzer({ language:
    'en' })
        const { dfData, username } = req.body
      //check if searched user is in the data
        const df = new DataFrame(dfData)
        let removeUserRow = df.query({
          column: "users",
          is: "!=",
          to: username
        })
        //filter rows with tweet length <=1
        let filterByLength = removeUserRow.query({
          column: "length",
          is: ">",
          to: 1
    ```

```
      })
.  .  .  .  .  .
      }
```

In the preceding code, we first instantiated `SentimentAnalyzer`, and then set its language configuration to English (en). We then obtained `dfData` and `username` from the request body.

To analyze and create insights from the data, we only want to consider the user's interaction with others and not themselves; that is, we don't want to consider tweets where the user replied to themselves. Hence, we filter out the user's replies from the DataFrame generated from `dfData`.

Sometimes, a tweet will just contain a hashtag or reference a user with the @ sign. However, we removed the hashtags and @ signs during our cleaning process earlier, which will result in some tweets ending up not containing any text. Therefore, we will create a new `filterByLength` DataFrame that will contain non-empty text:

1.  We will move ahead to creating an object that will contain the overall sentiment count for the user's data and is sent to the frontend whenever we make a call to the API:

```
let data = {
   positive: 0,
   negative: 0,
   neutral: 0
}
let sent = filterByLength["text"].values
for (let i in sent) {
   const getSent = await sentiment.getSentiment(sent[i])
   if (getSent.vote === "negative") {
     data.negative += 1
   } else if (getSent.vote === "positive") {
     data.positive += 1
   } else {
     data.neutral += 1
   }
}
res.status(200).json(data)
```

In the preceding code, we create object data to store the overall sentiment analysis. Since sentiment analysis will only be performed on the text data from the `filterByLength` DataFrame, we extract the text column values.

2.  We then loop through the text column values extracted and pass them into `sentiment.getSentiment`. For each piece of text passed into `sentiment.getSentiment`, the following types of objects are returned:

```
{
    score: 2.593,
    numWords: 36,
    numHits: 8,
    average: 0.07202777777777777,
    type: 'senticon',
    locale: 'en',
    vote: 'positive'
}
```

For our use case, we only need the key value of `vote`. Hence, we check whether the value of `vote` for a text is `negative`, `positive`, or `neutral`, and we then increment the count of each of these keys in a data object.

Hence, whenever a call is made to `/api/nlp`, we should receive the following response, for example:

```
{
    positive: 20,
    negative: 12,
    neutral: 40
}
```

In this section, we saw how we can create APIs in Next.js and, more importantly, we saw how it is convenient to use Danfo.js in the backend. In the next section, we will implement the frontend part of the app.

# Building the frontend

For our frontend design, we will use the default UI that comes with Next.js, as shown in *Figure 12.3*. We will implement the following set of components for our frontend:

- The `Search` component: Creates a search box to search for Twitter users.

- The `ValueCount` component: Obtains the count of unique values and plots it using a bar chart or pie chart.

- The `Plot` component: This component is used to plot our sentiment analysis in the form of a bar chart.

- The `Table` component: This is used to display the obtained user data in table form.

In the following sections, we'll implement the preceding list of components. Let's get started by implementing the `Search` component.

## Creating the Search component

The `Search` component is the main component for setting the app in action. It provides the input field in which a Twitter user's name can be inputted and then searched for. The `search` component enables us to make a call to the two APIs created: `/api/tweet` and `/api/nlp`.

In our `twitterdashboard` project directory, let's create a directory called `components`, and in that directory, we'll create a JavaScript file named `Search.js`.

In `Search.js`, let's input the following code:

```
import React from 'react'
export default function Search({ inputRef, handleKeyEvent,
handleSubmit }) {
   return (
      <div className='border-2 flex justify-between p-2 rounded-
md   md:p-4'>
         <input id='searchInput'
            type='text'
            placeholder='Search twitter user'
            className='focus:outline-none'
            ref={inputRef}
            onKeyPress={handleKeyEvent}
         />
         <button className='focus:outline-none'
```

```
                onClick={() => { handleSubmit() }}>
                <img src="/search.svg" />
        </button>
    </div>
    )
}
```

In the preceding code, we created a `Search` function with the following set of props:

- `inputRef`: This prop is obtained from the `useRef` React Hook. It will be used to track the current value of the search input field.

- `handleKeyEvent`: This is an event function that will be passed to the search input field to enable search by just pressing the *Enter* key.

- `handleSubmit`: This is a function that will be activated anytime you click the search button. The `handleSubmit` function is responsible for making a call to our API.

Let's move on to `/pages/index.js` to update the file by importing the `Search` component and creating the preceding list of required props based on the following steps:

1.  First, we will import React, React Hooks, and the `Search` component:

    ```
    import React, { useRef, useState } from 'react'
    import Search from '../components/Search'
    ```

2.  Then, we will create some sets of state:

    ```
    let [data, setData] = useState() // store tweet data from
    /api/tweet
    let [user, setUser] = useState() // store twitter
    usersname
    let [dataNlp, setDataNlp] = useState() // store data from
    /api/nlp
    let inputRef = useRef() // monitor the current value of
    search input field
    ```

3.  We will then create the `handleSubmit` function to make a call to our API and update the state data:

    ```
    const handleSubmit = async () => {
        const res = await fetch(
            '/api/tweet',
    ```

```
        {
            body: JSON.stringify({
            username: inputRef.current.value
            }),
            headers: {
            'Content-Type': 'application/json'
            },
            method: 'POST'
            }
    )
        const result = await res.json()
.  .  .  .  .  .  .  .
    }
```

First, in `handleSubmit`, we make a call to the `/api/tweet` API to obtain the user's data. In the `fetch` function, we obtain the current value of the `inputRef. current.value` search field and convert it into a JSON object, which is passed into the request body. A variable result is then used to obtain the JSON data from the API.

4. We further update the `handleSubmit` function to fetch data from `/api/nlp`:

```
const resSentiment = await fetch(
        '/api/nlp',
        {
            body: JSON.stringify({
            username: inputRef.current.value,
            dfData: result
            }),
            headers: {
            'Content-Type': 'application/json'
            },
            method: 'POST'
        },
    )
    const sentData = await resSentiment.json()
```

The preceding code is the same as that in *Step 3*. The only difference is that we make a call to the `/api/nlp` API and then pass the result data from *Step 3* and the username obtained from the search input field into the request body.

5.  We then update the following states in `handleSubmit`:

```
setDataNlp(sentData)
setUser(inputRef.current.value)
setData(result)
```

6.  Next, we will create the `handleKeyEvent` function to enable search by pressing the *Enter* key:

```
const handleKeyEvent = async (event) => {
    if (event.key === 'Enter') {
      await handleSubmit()
    }
  }
```

In the preceding code, we check whether the keypress is an *Enter* key, and if it is, we make a call to the `handleSubmit` function.

7.  Finally, we make a call to our `Search` component:

```
<Search inputRef={inputRef}
handleKeyEvent={handleKeyEvent}
handleSubmit={handleSubmit} />
```

Remember, we said that we'll be using Next.js's default UI. Hence, in `index.js`, let's convert `Welcome to Next.js` to `Welcome to Twitter Dashboard`.

After updating `index.js`, you can check your browser for the update at `http://localhost:3000/`. You'll see the following changes:

Figure 12.4 – index.js updated with the search component

The Search component is implemented and infused into the main app, and all the required state data can easily be updated by the Search component. Let's go ahead and implement the ValueCounts component.

# Creating the ValueCounts component

We will be creating a simple analysis for the obtained data from /api/tweet. This analysis involves checking the number of times the unique values in a column exist. We'll obtain the value count of the source column and the users column.

The value count from the source column tells us the device used by other Twitter users to interact with our searched user. The value count from the users column tells us the users who interact most with our searched user.

> **Note**
>
> The code used here is copied from the *Implementing the chart component* section in *Chapter 8, Creating a No-Code Data Analysis/Handling System*. The code for this section can be obtained here: https://github.com/PacktPublishing/Building-Data-Driven-Applications-with-Danfo.js-/blob/main/Chapter12/components/ValueCounts.js. Most of the code won't be explained in detail here.

Let's go to the components/ directory and create a file named ValueCounts.js, and then update it with the following steps:

1.  First, we import the necessary modules:

    ```
    import React from "react"
    import { DataFrame } from 'danfojs/src/core/frame'
    import { Pie as PieChart } from "react-chartjs-2";
    import { Bar as BarChart } from 'react-chartjs-2';
    ```

2.  Then, we create a function called ValueCounts:

    ```
    export default function ValueCounts({ data, column,
    username, type }) {


    }
    ```

The function accepts the following props:

a) data: This is the data from /api/tweet.

b) column: The name of the column from which we want to obtain the value count.

c) username: The inputted username from the search field.

d) type: The type of chart we want to plot.

3.   Next, we update the ValueCounts function:

```
const df = new DataFrame(data)
const removeUserData = df.query({
   column: "users",
   is: "!=",
   to: username
})
const countsSeries = removeUserData[column].value_
counts()
const labels = countsSeries.index
const values = countsSeries.values
```

In the preceding code, we first create a DataFrame from the data and then filter out the rows containing the searched user since we don't want tweets where the user is interacting with themself. We then extract the value_counts value from the passed-in column. From the created countSeries variable, we generate our labels and values, which we will use for plotting our chart.

4.   We then create a chart data variable called dataChart, which will be in the format accepted by the chart component:

```
const dataChart = {
    labels: labels,
    datasets: [{
       . . . .
       data: values,
    }]
};
```

The dataChart object contains the labels and values created in *Step 3*.

5.  We create a conditional rendering to check the type of chart to plot:

```
if (type === "BarChart") {
    return (
        <div className="max-w-md">
        <BarChart data={dataChart} options={options}
width="100" height="100" />
        </div>
    )
} else {
    return (<div className="max-w-md">
            <PieChart data={dataChart} options={options}
width="100" height="100" />
        </div>)
}
```

The `ValueCounts` component is set.

We can now import the `ValueCounts` component into `index.js` using the following steps:

1.  We import `ValueCounts`:

```
import dynamic from 'next/dynamic'
const DynamicValueCounts = dynamic(
    () => import('../components/ValueCounts'),
    { ssr: false }
)
```

The way we import `ValueCounts` is different from the way we import the `Search` component. This is because in `ValueCounts`, we use some core browser-specific tools in TensorFlow.js, which is required in Danfo.js. Hence, we need to prevent Next.js from rendering the component from the server to prevent an error.

To prevent Next.js from rendering a component from the server, we use `next/dynamic`, and we then wrap the component to be imported in the `dynamic` function, as well as setting the `ssr` key to `false`.

---

**Note**

To find out more about `next/dynamic`, check out `https://nextjs.org/docs/advanced-features/dynamic-import`.

2.  We make a call to the `ValueCounts` component, which is now named `DynamicValueCounts`:

    ```
    {typeof data != "undefined" && <DynamicValueCounts
    data={data} column={"source"} type={"PieChart"} />}
    ```

    We check whether the state data is undefined and that the user's data has been fetched from `/api/tweet`. If so, we render the `ValueCounts` component for the `source` column.

3.  Let's also add `ValueCounts` for the `users` column:

    ```
    {typeof data != "undefined" && <DynamicValueCounts
    data={data} column={"users"} username={user}
    type={"BarChart"} />}
    ```

    We specify `BarChart` for the user's `ValueCounts` chart and `PieChart` for the `ValueCounts` source.

The following shows the display of `ValueCounts` for the source and user interactions whenever a user is searched:

# Welcome to Twitter Dashboard!

reactjs

## Tweet source

Display the device for all the tweet user is mentioned in.

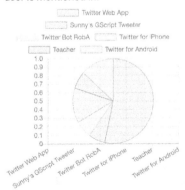

## User Interactions

A chart of users that interact with the given twitter username.

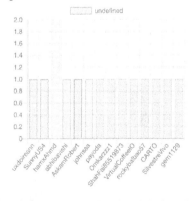

Figure 12.5 – ValueCounts chart result for the source and user columns

The value count is done and working okay. Let's move on to create a plot for our sentiment analysis data obtained from /api/nlp.

# Creating a plot component for sentiment analysis

When a user is searched by using the search field, the sentiData state is updated to contain the sentiment data from /api/nlp. In this section, we will create a Plot component for the data.

Let's create a Plot.js file in the components/ directory and update it with the following steps:

1.  First, we import the required modules:

    ```
    import React from "react"
    import { Bar as BarChart } from 'react-chartjs-2';
    ```

2.  We then create a Plot function to plot a chart for sentiment data:

    ```
    export default function Plot({ data }) {
      const dataChart = {
        labels: Object.keys(data),
        datasets: [{
          . . . . . . . .
          data: Object.values(data),
        }]
      };
      return (
        <div className="max-w-md">
          <BarChart data={dataChart} options={options}
    width="100" height="100" />
        </div>
      )
    }
    ```

The function accepts a data prop. We then create a dataChart object containing the format for the chart component. We specify the chart label by obtaining the keys in the data props and also specify the value of key data in dataChart by obtaining the values of the data props. The dataChart object is passed into the BarChart component. The Plot component is now created for the sentiment analysis chart.

3.  The next step is to import the `Plot` component inside `index.js` and make a call to it:

```
import Plot from '../components/Plot'

. . . . . .


{typeof dataNlp != "undefined" && <Plot data={dataNlp}
/>}
```

With the preceding update in `index.js`, we should see the following chart for sentiment analysis whenever we search for a user:

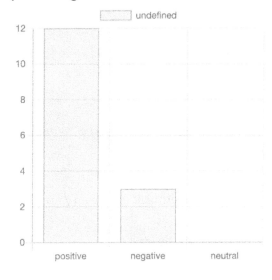

Figure 12.6 – Sentiment analysis chart

The sentiment analysis is done and fully integrated into `index.js`. Let's move on to create the `Table` component to display our user data.

## Creating a Table component

We will be implementing a `Table` component to display the data obtained.

> **Note**
>
> The table implementation is the same as for the DataTable implementation created in *Chapter 8, Creating a No-Code Data Analysis/Handling System*. For a better explanation of the code, kindly check out *Chapter 8, Creating a No-Code Data Analysis/Handling System*.

Let's create a `Table.js` file in the `components/` directory and update the file with the following steps:

1. We import the necessary modules:

```
import React from "react";
import ReactTable from 'react-table-v6'
import { DataFrame } from 'danfojs/src/core/frame'
import 'react-table-v6/react-table.css'
```

2. We create a function called `Table`:

```
export default function DataTable({ dfData, username }) {

}
```

The function takes in `dfData` (sentiment data from `/api/nlp`) and `username` (from the search field) as props.

3. We update the function with the following code:

```
const df = new DataFrame(dfData)
const removeUserData = df.query({
  column: "users",
  is: "!=",
  to: username
})
const columns = removeUserData.columns
const values = removeUserData.values
```

We create a DataFrame from `dfData`, filter out the rows containing the user's tweets, and then extract the column names and the values of the DataFrame.

4.  We then format this column in the format accepted by `ReactTable`:

```
const dataColumns = columns.map((val, index) => {
    return {
      Header: val,
      accessor: val,
      Cell: (props) => (
        <div className={val || ''}>
        <span>{props.value}</span>
        </div>
      ),
      . . . . . .
    }
  });
```

5.  We also format the values to be accepted by `ReactTable`:

```
const data = values.map(val => {
    let rows_data = {}
    val.forEach((val2, index) => {
      let col = columns[index];
      rows_data[col] = val2;
    })
    return rows_data;
  })
```

Once again, the code in *Steps 4, 5,* and *6* is explained in detail in *Chapter 8, Creating a No-Code Data Analysis/Handling System.*

6.  We then make a call to the `ReactTable` component and pass in `dataColumns` and `data`:

```
<ReactTable
  data={data}
  columns={dataColumns}
  getTheadThProps={() => {
    return { style: { wordWrap: 'break-word', whiteSpace:
'initial' } }
```

```
    }}
    showPageJump={true}
    showPagination={true}
    defaultPageSize={10}
    showPageSizeOptions={true}
    minRows={10}
/>
```

The `table` component is completed; the next step is to import the component in Next.js and then make a call to the component.

Note that since we are using the web version of Danfo.js, we need to load this component with `next/dynamic` to prevent the app from crashing:

```
const Table = dynamic(
    () => import('../components/Table'),
    { ssr: false }
)
. . . . . .
{typeof data != "undefined" && <Table dfData={data}
username={user} />}
```

In the preceding code, we import the `Table` component dynamically and also instantiate the `Table` component inside and pass in the `dfData` and `username` prop values.

If you switch over to your browser and go to the project's `localhost` port, you should see the full updated app, as shown in the following screenshot:

## User Data

A table showing the extracted data for the given twitter username.

| text | length | date | s |
|------|--------|------|---|
| learning redu | 19 | Fri Jul 16 10:4 | Twit |
| password cra | 3 | Fri Jul 16 10:3 | Sun |
| thats great | 2 | Fri Jul 16 10:' | Twit |
| to speed up r | 40 | Fri Jul 16 10:( | Twit |
| absolutely 7 ', | 19 | Fri Jul 16 10:( | Twit |
| absolutely 7 ', | 34 | Fri Jul 16 10:( | Twit |
| our headless | 35 | Fri Jul 16 10:( | Twit |
| thank you for | 24 | Fri Jul 16 09: | Tea |
| thank you be | 46 | Fri Jul 16 09: | Twit |
| password cra | 3 | Fri Jul 16 09: | Twit |

Page    1    of 2

Previous                          Next

10 rows    ∨

Figure 12.7 – Extracted user data

The final result of the app should look like the following:

# Welcome to Twitter Dashboard!

reactjs

### Tweet source

Display the device for all the tweet user is mentioned in.

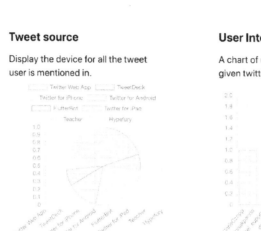

### User Interactions

A chart of users that interact with the given twitter username.

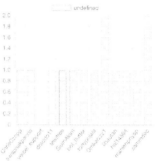

### User Data

A table showing the extracted data for the given twitter username.

| text | length | date | s |
|---|---|---|---|
| writing after | 3 | Sun Jul 11 06 | Hyp |
| how many pe | 36 | Sun Jul 11 05 | Twi |
| thank you so | 4 | Sun Jul 11 05 | Twi |
| ah this is real | 18 | Sun Jul 11 04 | Twi |
| oh really that | 18 | Sun Jul 11 04 | Twi |

### Sentiment Analysis

A chart showing the overall sentiment percentage for all the tweet.

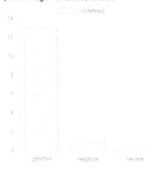

Previous    Page    2    of 2    Next

10 rows ⌄

Figure 12.8 – Twitter user dashboard

In this section, we built different components for the frontend implementation. We saw how to make use of Danfo.js with Next.js and also got to know about loading components with `next/dynamic`.

## Summary

In this chapter, we saw how to make use of Next.js to build a fast full stack app. We saw how to make use of Danfo.js nodes in the backend and we also used JavaScript packages such as twit.js and `nlp-node` to obtain Twitter data and perform sentiment analysis.

We also saw how to easily infuse Danfo.js with Next.js and how to prevent errors by loading components with `next/dynamic`.

The goal of the chapter was to enable you to see how you can easily use Danfo.js to build a full stack (backend and frontend) data-driven app, and I believe this was well achieved in this chapter.

I believe we've covered a lot in this book, starting from the introduction to Danfo.js to building a no-code environment with Danfo.js, to building a recommendation system and Twitter analytics dashboard. From the various use cases of Danfo.js with various JavaScript frameworks, we were able to build an analytics platform and machine learning-driven web apps.

We've come to the end of the book and I believe we are now well equipped with skills to include data analytics and machine learning in our next web app and also contribute to Danfo.js.

# 13
# Appendix: Essential JavaScript Concepts

Welcome to the final chapter of this book. We have placed this chapter at the end of this book so as not to bore the experienced JavaScript developers with introductory concepts. This chapter should be read by developers who want a refresher on the basics of JavaScript before attempting to use `Danfo.js`.

By the end of this chapter, you'll understand basic concepts of JavaScript that are essential to building applications in the language. You'll learn about data types, conditional branching and looping constructs, and JavaScript functions.

Specifically, we'll cover the following concepts:

- Quick overview of JavaScript
- Understanding the fundamentals of JavaScript

# Technical requirements

We will be using the developer console for all code examples in this chapter. To run any code snippet in your default browser, you need to open the developer console. The commands for opening the console provided for various browsers are shown in the following list:

- **Google Chrome**: To open the developer console in Google Chrome, open the **Chrome** menu from the upper-right-hand corner of the browser window and select **More Tools | Developer Tools**. You can also use the *Option + Command + J* shortcut on macOS, or *Shift + Ctrl + J* on Windows/Linux.

- **Microsoft Edge**: In Microsoft Edge, open the **Edge** menu in the upper-right-hand corner of the browser window and select **F12 Developer Tools**, or you can just press *F12* to open it.

- **Mozilla Firefox**: In Mozilla Firefox, click on the **Firefox** menu in the upper-right-hand corner of the browser and select **Web Developer | Browser Console**. You can also use the *Shift + ⌘ + J* shortcut on macOS or *Shift + Ctrl + J* on Windows/Linux.

- **Apple Safari**: In Safari, you'll first need to enable the **Developer** menu in your browser settings. To do that, open Safari's preferences (*Safari Menu* | **Preferences**), select the **Advanced** tab, and then enable the **Developer** menu. Once that menu is enabled, you will find the developer console by clicking on **Develop | Show JavaScript Console**. You can also use the *Option + ⌘ + C* shortcut.

Once your developer console is open, depending on your browser, you will see a console similar to the one shown here:

Figure 13.1 – Chrome browser developer console

With your console open, you're ready to start writing and testing JavaScript code. In the next section, we'll quickly go over some important concepts of the JavaScript language.

# Quick overview of JavaScript

According to the *Stack Overflow 2020 Developer Survey* (`https://stackoverflow.blog/2020/05/27/2020-stack-overflow-developer-survey-results/`), **JavaScript** (also referred to as **JS**) is the most common programming language in the world, with approximately 70% of developers using it for one task or another. This statistic is not surprising as JavaScript had been the most popular language for a good number of years before the survey was carried out. There are many reasons why this is so, and we will list some of them here:

- It runs in the most common and readily available platform—the browser.
- Numerous useful frameworks such as Node.js, React, and Angular are built around it.
- It is versatile—that is, it can be used for both frontend and backend applications. For example, you can use JavaScript libraries such as React, Vue, and Angular to build great **user interfaces** (**UIs**), while you can use server-side packages such as Node.jS and Deno to build efficient backend/server-side applications.
- It can be used for **Internet of Things** (**IoT**) and cross-platform mobile applications.

JavaScript was initially created to be a browser-only language but has quickly evolved to be used almost everywhere, from frontend applications to backend applications with Node.js, to IoT applications, and—more recently—in the data science/**machine learning** (**ML**) field.

> **Note**
> Do not confuse JavaScript with the Java programming language (`https://en.wikipedia.org/wiki/Java_(programming_language)`). Although the names may be similar, they are very much different in terms of use, syntax, and even semantics.

JavaScript is dynamic and event-driven and thus is the subject of numerous concerns, especially for programmers coming from other languages. This has led to the creation of languages that can be directly transpiled to JavaScript. Some of these languages are **TypeScript** (a language with strict data typing developed by Microsoft), **Dart** (a standalone language developed by Google), **CoffeeScript**, and so on.

# Understanding the fundamentals of JavaScript

In this section, as a memory refresher, we'll quickly go over the basic concepts of modern JavaScript. If you are familiar with JavaScript, then you can skip this section.

As we have repeatedly mentioned, JavaScript can be used for both frontend and server-side scripting, so there is syntax or features that are particular to each environment— for instance, browser-side JavaScript does not have access to filesystems such as Node.js because of security reasons. So, in this section, we'll introduce concepts that can work in both environments/any environment.

## Declaring variables

**Variables** are named storage for data. They can be used to store data that JavaScript can manipulate or work with. In JavaScript, you can define variables using two main keywords, `const` and `let`, although in older scripts, you'll find the `var` keyword used for declaring variables. Using `var` for variable declaration is generally discouraged and should only be used rarely. In a later section, we will discuss some reasons why it is encouraged to use `let` instead of `var` in modern JavaScript.

The following statement declares a variable using the `let` keyword:

```
let name;
let price = 200;
let message, progress, id;

message = "This is a new message";
progress = 80;
id = "gh12345";
```

The next keyword, `const`, can be used to declare a constant variable—that is, a variable whose reference cannot be changed during the runtime of an application.

Some examples of `const` declarations are shown in the following code snippet:

```
const DB_NAME;
const LAST_NAME = "Williams";
const TEXT_COLOR = "#B00";
const TITLE_COLOR = "#F00";
```

> **Important note**
>
> Using uppercase variable names when declaring constants is a common and widely encouraged practice, not just in JavaScript but in many programming languages.

# Data types

JavaScript supports eight basic types of data. Data types of declared variables are automatically inferred at runtime by the JavaScript compiler because JavaScript is a dynamically typed language. The eight supported data types are **number**, **string**, **Boolean**, **object**, **BigInt**, **undefined**, **null**, and **symbol**.

## Number

The number type can represent integers, floating-point numbers, infinity, -infinity, and **Not a Number** (**NaN**). Number types support numerous mathematical operations such as addition, subtraction, multiplication, division, and so on. In the following code snippet, we define some variables of type number:

```
let num1 = 10; //integer
let num2 = 20.234; //floating point number
console.log(num1 + num2); //outputs 30.234
```

> **Important note**
>
> infinity and -infinity represent mathematical positive and negative infinity (∞). These values are bigger than any number. You can see an example here:
>
> ```
> console.log(1 / 0); // Infinity
> ```
>
> The NaN sub-types represent an error in a mathematical operation with numbers—for instance, trying to divide a string with a number, as illustrated here:
>
> ```
> console.log("girl" / 2); // NaN
> ```

## String

Strings in JavaScript represent texts and must be surrounded with single (' '), double
(" "), or backtick (' ') quotes. You can see some examples of strings in the following
code snippet:

```
let name = "John Doe";
let msg = 'I am on my way';
let address = '10 Slow ave, NY';
```

The choice of double or single quotes is based on preference as they both perform the
same function. Backticks, on the other hand, have more functionalities than basic quotes.
They allow you to embed variables and expressions into strings easily, using the ${...}
template literal syntax. For example, in the following code snippet, we show how to easily
embed name, address, and dog_counts variables in a new variable message:

```
let name = "John Doe";
let address = '10 Slow ave, NY';
let dog_counts = 10;
let msg = '${name} lives in ${address} and has ${dog_counts}
dogs';
console.log(msg)


//outputs
John Doe lives in 10 Slow ave, NY and has 10 dogs
```

> **Tip**
>
> A string data type has many built-in functions for manipulating it. You can
> find many of these functions on the **Mozilla Developer Network** (**MDN**)
> documentation page (https://developer.mozilla.org/en-
> US/docs/Web/JavaScript/Reference/Global_Objects/
> String).

## Boolean

The Boolean data type is a logical type. It has only two values: `true` and `false`. Boolean values are mostly used in comparison operations and for storing *yes*/*no* values. Let's look at the following example:

```
let pageClicked = false; // No, the page was not clicked
let pageOpened = true; // Yes, the page was opened

if (5 > 1) { //Evaluates to the Boolean value true
  console.log("Yes! It is.")
}
```

Next, let's talk about the object type.

## Object

The object in JavaScript is a very special type. It is perhaps the most important type in JavaScript and is used in almost every aspect of the language. The object data type is the only non-primitive type in JavaScript. It can store keyed collections of different types of data, including itself.

> **Tip**
> Primitive types in programming languages are the most basic types and generally store a single kind of value. Non-primitive types, on the other hand, can store more than one type of value and can also be extended to perform other functions.

There are two ways for creating objects, outlined here:

- The first and most common way is to use curly brackets with an optional list of key-value properties, as illustrated here:

```
let page = {}
```

- And a not so common way is by using object constructor syntax, as follows:

```
let page = new Object()
```

Objects can be created and initialized in the same step, as shown in the following code snippet:

```
let page = {
  title: 'Home page',
```

```
    Id: 23422,
    text: 'This is the first page in my website',
    rank: 32.09,
    "closed": false,
    owner: {
      name: 'John Doe',
      level: 'admin',
      lastLoggedIn: 'Aug 5th',
    },
}
```

Notice that an object page can store numerous types of data, even other objects (as the owner). This shows that objects can be nested to contain more and more objects.

To access object values, you can use square brackets, like this:

```
console.log(page["closed"])
//outputs false
```

Alternatively, you can use dot `page.rank` notation, like this:

```
console.log(page.rank)
```

This gives the following output:

```
//outputs 32.09
```

In *Chapter 1*, *Overview of Modern JavaScript*, you learned about some important properties of objects that make them very powerful.

# Conditional branching and loops

Conditional branching and loops are important aspects of any programming language, and JavaScript provides them as well. To perform different actions based on some conditions, you can use `if`, `if..else`, conditional/ternary operators, and `switch` syntax. Besides conditional branching, you may want to perform repeated actions a specific number of times. This is where looping comes into play. JavaScript provides looping constructs such as `while`, `do...while`, `for..of`, `for..in`, and traditional `for` loops. We'll briefly cover each of these statements in the following section.

## if statement

An `if` statement evaluates an expression, a condition, or a piece of code in the parentheses and will only execute the statement if it is `true`. This is a one-way conditional that only runs when it is true. You can see an example of this in the following code snippet:

```
let age = 25
if (age >= 18){
   console.log("You're an adult!")  //will run since age is
greater than 18
}
//outputs
You're an adult!
```

## if...else statement

An `if...else` statement provides an extra block that runs when the initial condition becomes `false`. Here is an example of this:

```
let age = 15
if (age > 18){
   console.log("You're an adult!")
}else{
   console.log("You're still a teenager!")
}
//outputs
You're still a teenager!
```

## Conditional/ternary operator

A ternary operator is a shorter and more concise way to write an `if..else` statement. It is mostly used in assigning variables based on two conditions. The syntax for ternary operators is illustrated here:

```
let result = condition ? val1 : val2;
```

If the condition is true, then `val1` gets executed and assigned to `result`, else `val2` becomes assigned. An example of using this is shown in the following code snippet:

```
let age = 20
let isAdult = age > 18 ? true : false; //age is greater than
```

```
18, so true is returned
console.log(isAdult)
//outputs
true
```

## switch statement

A switch statement can be used to replace multiple if...else statements in a more concise and readable way. The syntax is shown here:

```
switch(value) {
  case 'option1':  // if (value === 'option1')
    ...
    [break]
  case 'option2':  // if (value === 'option2')
    ...
    [break]
 case 'option3':  // if (value === 'option3')
    ...
    [break]
  default:
    ...
    [break]
}
```

The variable or expression in the switch parentheses is checked against each case statement, and the corresponding code is executed if the condition is true. The default statement gets executed when every other condition fails. The following code snippet shows an example of using a switch statement:

```
let value = 4
switch (value) {
  case 1:
    console.log( 'You selected 1' );
    break;
  case 2:
    console.log( 'You selected 2' );
    break;
  case 4:
```

```
        console.log( 'You selected 4!' ); //gets executed
        break;
    default:
        console.log( "No such value" );
}
```

```
//outputs
"You selected 4!"
```

A break statement after each case is important in order to stop the execution of succeeding cases. If you neglect the break statement, then the succeeding cases will also be executed.

## while loop

A while loop will execute a block of code repeatedly as long as the condition is true. The syntax is shown in the following code snippet:

```
while (condition) {
    // loop body
}
```

Let's look at the following example of using a while loop:

```
let year = 2015;
while (year <= 2020) { ⬎
    console.log( year );
    year++
}
//outputs
2015
2016
2017
2018
2019
2020
```

The preceding code will continually print out the year as long as it is less than or equal to 2020. Note the year++ part. This is important in order to break the while loop at some point.

## do...while loop

A do...while loop is very similar to a while loop, with just one little difference— the body is executed at least once before the condition is tested. The syntax is shown here:

```
do {
    // loop body
} while (condition);
```

Let's look at the following example of a do...while loop in action:

```
let year = 2015;
do {
    console.log( year );
    year++;
} while (year <= 2020);
```

A do...while loop is important if you need to execute a piece of code at least once before a conditional check is carried out.

> **Important note**
> Always remember to set a condition that breaks a loop at some point, else your loops will execute forever—theoretically, though, the browser and server side will stop such loops after a certain time.

## for loop

A for loop is the most popular looping construct in JavaScript. The syntax is shown here:

```
for (initialization; condition; step) {
    // loop body
}
```

The syntax has three important parts, outlined as follows:

- **The initialization part**: This gets executed before the loop is started. You will mostly initialize an iterator variable here—for instance, (`let i = 0`).

- **The condition part**: This code is checked before each loop interaction. If this is false, then the loop is stopped.

- **The step part**: The step is important because it will generally increase a counter or variable that will be tested.

An example of using a `for` loop is given here:

```
for (let i=0; i <=5; i++){
    console.log(i)
}
//outputs
0
1
2
3
4
5
```

> **Note**
>
> Notice that we used a post-incremental operator (`i++`) in the step part of the `for` loop? This is pretty standard, and this is just shorthand for `i = i + 1`.

## for...of and for...in loops

`for...of` and `for...in` loops are variants of a `for` loop for iterating over iterables. **Iterables** are special data types such as objects, arrays, maps, and strings that have an iterable property—in other words, the properties or values can be looped through.

A `for...of` loop is mostly used for iterating over objects such as arrays and strings. The syntax is shown here:

```
for (let val of iterable) {
    //loop body
}
```

Let's look at the following example of using a `for...of` loop:

```
let animals = [ "bear", "lion", "giraffe", "jaguar" ];
// Print out each type of animal
for (let animal of animals) {
    console.log(animal);
}
//outputs
"bear"
"lion"
"giraffe"
"jaguar"
```

A `for...in` loop is used to loop over objects. The syntax is very similar to a `for...of` loop, as shown in the following code snippet:

```
for (let val in iterable) {
    //loop body
}
```

A `for...in` loop should be used only with enumerables such as objects and not with iterables such as arrays.

An example of using a `for...in` loop is shown here:

```
let user = {
  name: "Hannah Montana",
  age: 24,
  level: 2,
};
for (key in user) {
    console.log(user[key]);
}
//outputs
"Hannah Montana"
24
2
```

> **Further reading**
>
> `for...in` and `for...of` are looping constructs that are used to iterate over data structures. The main difference between them is the data structure they iterate over. `for...in` iterates over all enumerable property keys of an object, while `for...of` iterates over the values of an iterable object.

# JavaScript functions

In this section, we will learn about JavaScript functions. **Functions** are very important when we need to write cleaner code that reuses existing blocks in numerous parts of our script. Since the beginning of this chapter, we've been calling the `console.log` built-in function whenever we need to print text to the console. This shows how important functions can be, as a single function, once defined, can be called in any way and any number of times to perform the same action.

A function in JavaScript typically looks like this:

```javascript
function sayHi() {
   console.log('Hello World!')
}
sayHi()
//output
"Hello World!"
```

A function can also take a list of comma-separated parameters inside parentheses, which can be used to perform computation. For instance, a function with some parameters will typically look like this:

```javascript
function sayHi(name, message) {
   console.log('This is a message from ${name}. The message read
thus: ${message}')
}

sayHi("Jenny", "How are you doing?")
//output
"This is a message from Jenny. The message read thus: How are
you doing?"
```

Some important things to note about functions are listed here:

- Variables declared inside a function are local to that function and can only be accessed inside the function. Here's an example of this:

```
function sayHi() {
    let name = 'Jane'
    console.log(name)
}

sayHi()
//output
"Jane"
console.log(name) //throws error: ReferenceError: name is
not defined
```

- A function can access global variables. Global variables are variables declared outside the function scope and are available to every code block. You can see an example of this here:

```
let name = 'Jenny'
function sayHi() {
    console.log(name)
}
sayHi()
//output
"Jenny"
```

- Functions can return values. This is important when a function performs some computation and the result needs to be used elsewhere. The functions we have been working with up till now mostly return nothing/undefined and they are referred to as void functions. Let's look at the following example of a function returning a value:

```
function add(num1, num2) {
    let sum = num1 + num2
    return sum
}
let result = add(25, 30)
```

```
console.log(result)

//output
55    •
```

---

**Tip**

The `return` statement in a function can only return a single value at a time. In order to return more than one value, you can return the result as a JavaScript object. You can see an example of this here:

```
function add(num1, num2) {
  let sum = num1 + num2
  return {sum: sum, funcName: "add"}}
let result = add(25, 30)
  console.log(result)
//output
Object {  funcName: "add",   sum: 55}
```

---

## Callback functions

Callbacks are very important in JavaScript. A callback is a function passed into another function as an argument. This argument is called to perform some action based on some conditions in the outer function. Let's look at the following example to better understand callbacks:

```
function showValue(value) {
  console.log(value)
}
function err() {
  console.log("Wrong value specified!!")
}
```

In the preceding code example, we created two callback functions, `showValue` and `error`. You can see these in use in the following code snippet:

```
function printValues(value, showValue, error) {
  if (value > 0){
    showValue(value)
```

```
  }else{
    error()
  }
}
```

The `showValue` callback function prints the argument passed to it to the console, while the `error` callback function prints an error message. The callback functions can now be used, as shown in the following code snippet:

```
printValues(20, showValue, err);
//outputs 20


printValues(0, showValue, err);
//outputs "Wrong value specified!!"
```

The `printValues` function first tests if the value is greater than 0, and then either calls the `showValue` or the `error` callback function.

Callbacks don't just take in named functions. They can also take in anonymous functions; with this, it is possible to create *nested callbacks*.

Let's assume we have a `doSomething` function that does a particular task, and we would like to perform different operations before the task is completed. That means we can pass a callback function to another callback function, thereby creating nested callbacks, as shown in the following block of code:

```
doSomething((((((()=>{
//dosomething A
         })=>{
    //dosomething B
       })=>{
    //dosomething C
     })=>{
    //dosomething D
   })=>{
  //dosomething E
})
```

The preceding code is what we call a *callback hell*. This approach has lots of issues and becomes difficult to manage. In *Chapter 1*, An *Overview of Modern JavaScript*, we introduced a modern and more efficient way to work with callbacks by using a **application programming interface** (**API**). This is much cleaner and will eliminate many of the problems associated with callback hell.

> **Important note**
> JavaScript is asynchronous by default, thus long executing functions are queued and may never get executed before you need them. Callbacks are mostly used to continue code execution and ensure that the right result is returned.

# Summary

In this chapter, we introduced essential aspects of the JavaScript programming language. We started with some basic background that explained why JavaScript is the most popular language in the world today, as well as the various uses of JavaScript. Next, we looked at the fundamental concepts of the language, where we talked about declaring variables and also the eight data types available in JavaScript. Following that, we talked about branching and looping constructs in JavaScript and showed some examples of using them, and finally, we briefly discussed functions and classes in JavaScript.

# Other Books You May Enjoy

If you enjoyed this book, you may be interested in these other books by Packt:

**Hands-On Data Analysis with Pandas - Second Edition**

Stefanie Molin

ISBN: 978-1-80056-345-2

- Understand how data analysts and scientists gather and analyze data
- Perform data analysis and data wrangling using Python
- Combine, group, and aggregate data from multiple sources
- Create data visualizations with pandas, matplotlib, and seaborn
- Apply machine learning algorithms to identify patterns and make predictions
- Use Python data science libraries to analyze real-world datasets
- Solve common data representation and analysis problems using pandas
- Build Python scripts, modules, and packages for reusable analysis code

**Hands-On Machine Learning with scikit-learn and Scientific Python Toolkits**

Tarek Amr

ISBN: 978-1-83882-604-8

- Understand when to use supervised, unsupervised, or reinforcement learning algorithms

- Find out how to collect and prepare your data for machine learning tasks

- Tackle imbalanced data and optimize your algorithm for a bias or variance tradeoff

- Apply supervised and unsupervised algorithms to overcome various machine learning challenges

- Employ best practices for tuning your algorithm's hyper parametersAttack web and database servers to exfiltrate data

- Discover how to use neural networks for classification and regression

- Build, evaluate, and deploy your machine learning solutions to production

# Packt is searching for authors like you

If you're interested in becoming an author for Packt, please visit `authors.packtpub.com` and apply today. We have worked with thousands of developers and tech professionals, just like you, to help them share their insight with the global tech community. You can make a general application, apply for a specific hot topic that we are recruiting an author for, or submit your own idea.

# Share Your Thoughts

Now you've finished *Building Data-Driven Applications with Danfo.js*, we'd love to hear your thoughts! Scan the QR code below to go straight to the Amazon review page for this book and share your feedback or leave a review on the site that you purchased it from.

https://packt.link/r/1-801-07085-7

Your review is important to us and the tech community and will help us make sure we're delivering excellent quality content.

# Index

## Symbols

.apply method
  using 258-261

## A

accessors 159
aggregation, methods
  count() 263
  max() 263
  mean() 263
  min() 263
  std() 263
  sum() 263
Andrew Ng course 339
App.js
  ChartViz, and ChartPlane
    integrating into 322-325
apply function
  data transformation 134-136
Arithmetic component
  implementing 313-316
arithmetic computation, on tensors
  reference link 353

arithmetic operations
  about 110-114, 353
  applying, on tensors 354, 355
array methods
  about 14, 15
  Array.filter 17
  Array.includes 16
  Array.map 17
  Array.slice 17
  Array.splice 15, 16
arrays 87
arrow functions
  about 25-27
  using 25
artificial neural networks
  reference link 377
async/await 32-34

## B

Babel
  about 41
  reference link 48
  uses 41
  using, example projects 42-48

backend, Twitter Analysis dashboard
  text sentiment API, building  400-403
  Twitter API, building  396-400
bar charts
  creating  201-205
  creating, from multiple
      columns  242, 243
  creating, from Series  241
  creating, with Danfo.js  240
  reference link  205
basic charts
  bar charts, creating with
      Plotly.js  201-205
  bubble charts, creating with
      Plotly.js  205, 206
  creating, with Plotly.js  196
  reference link  207
  scatter plot, creating with
      Plotly.js  197-201
basic math operations
  applying, on tensors  356
Bayesian classifiers
  reference link  377
BigMart sales dataset  359
Boolean data type  427
box plots
  about  211
  creating, with Danfo.js  231
  creating, with Plotly.js  211, 212
  making, for multiple columns  233
  making, for Series  232
  making, with specific x and
      y values  235, 236
  reference link  215
Brain.js
  reference link  337
browser
  TensorFlow.js, setting up  343

bubble charts
  creating  205, 206
  reference link  206
buttons
  removing, from modebar  183, 184

C

callback functions  437, 438
callback hell  439
callbacks
  cleaning, with promises  29-32
Cartesian product  149
categorical data  201
cell
  about  63
  code cell  64
  Markdown cells  65
Chai
  unit testing  48
chart
  types  174
chart component
  implementing  317, 318
ChartPlane component
  implementing  318-321
ChartViz component
  implementing  321, 322
chart title
  configuring  190-192
ChartViz, and ChartPlane
  integrating, into App.js  322-325
class keyword  35, 36
closure  12, 13
clustering algorithms
  reference link  377
code cell  64
CoffeeScript  423

concatenation  153
conditional branching
  about  428
  if..else statement, using  429
  if statement, using  429
  switch statement, using  430, 431
  ternary operator, using  429
configuration options
  for plots  182
  reference link  182
content delivery network (CDN)  82, 175
convolutions, on tensors
  reference link  353
CoreAPI  348
Cs231n
  reference link  339
custom buttons
  adding, to modebar  185, 186
  properties  186

# D

D3.js
  URL  174
Danfo.js
  adding, to code  223, 224
  charts  222
  functions  101
  installing  84, 85
  methods  101
  need for  82-84
  reference link  84
  setting up, for plotting  222, 223
  used, for creating bar charts  240
  used, for creating box plots  231
  used, for creating histograms  237
  used, for creating line charts  225-229

used, for creating scatter plots  229-231
used, for creating violin plots  231
Danfo Notebook (Dnotebook) file  360
Dart  423
data
  grouping  247
  transforming  124
data aggregation
  about  245
  of grouped data  263
  on double-column grouping  264, 265
  on single-column grouping  263, 264
datacook
  reference link  337
data file formats
  loading  117-120
  methods  117
  working  117-120
data format  178-181
DataFrame
  about  85, 93-101
  transforming, into another
    file format  120
DataFrame data structures, methods
  .set_index() method  100
  .tensor  99
  .transpose() method  99
  .values  99
  head() method  94
  tail() method  94
DataFrame merge  147
DataFrame operation components
  Arithmetic component  295
  creating  294
  Describe component  295
  Describe component,
    implementing  295-297

Df2df component  295

Df2df component, implementing  306

Query component  295

Query component,
    implementing  301-303

DataFrames

  concatenating, along row
      axis (0)  154-156

  concatenating, along row axis (1)  157

  missing values, replacing  127-129

datasets

  combining  146

  combining, with DataFrame merge  147

  data concatenation  153

  inner merge, by single key  148, 149

  inner merge, on two keys  149

  outer merge, on single key  150

  outer merge, on two key  151, 152

  right and left merges  152

DataTable component

  about  282-284

  implementing  284-290

  integrating, into App.js  293, 294

data transformation

  about  124

  apply function, using  134-136

  DataFrames and Series,
      encoding  142-144

  filtering  137-139

  map function, using  131-134

  missing values, replacing  125

  one-hot encoding  144-146

  querying  137-139

  random sampling  139-141

data types, JavaScript

  about  425

  Boolean  427

  number  425

  object  427, 428

  string  426

date-time accessors  163-166

default locale

  changing  188, 189

delete keyword  19

Describe component

  implementing  295-297

  SidePlane, integrating into
      App.js  299-301

  SidePlane, setting up  297, 298

destructuring  7

Df2df component

  implementing  306-309

  interface, implementing  309-313

Django  82

Dnotebook

  about  60

  application areas  61

  installing  62, 63

  interactive computing  63

  setting up  62, 63

Dnotebook playground

  reference link  62

Dnotebook version 0.1.1  63

Document Object Model (DOM)  175

double-column grouped data

  iterating through  252-256

double-column grouping

  about  250-252

  data aggregation  264, 265

do...while loop  432

download plot options

  customizing  187

Draggable  272

duplicates

  removing  129-131

# E

embeddings
  about 378
  advantages 378, 379
evaluation metrics 333, 334

# F

fastai
  reference link 339
FastAPI 82
financial dataset
  downloading, for plotting 224, 225
five-dimensional (5D) tensor 352
Flask 82
for...in loop 433, 434
for loop 432, 433
for loop, syntax
  condition part 433
  initialization part 433
  step part 433
for...of loop 433, 434
four-dimensional (4D) tensor 352
frontend, Twitter Analysis dashboard
  building 404
  plot component, creating for
    sentiment analysis 412-414
  Search component, creating 404-408
  Table component, creating 414-419
  ValueCounts component,
    creating 408-412

# G

get_groups method
  using 256, 257
global function 11

global variable 11
groupby method, on real data
  application 265, 266
grouped box plots
  creating 214, 215
grouped data
  data aggregation 263
  iterating through 252
  standardizing 262
Grouplens
  URL 380
group operations 245

# H

histogram
  creating, from multiple
    columns 238, 239
  creating, from Series 237
  creating, with Danfo.js 237
histogram plots
  creating, with Plotly.js 207-211
  reference link 211
horizontal box plots
  creating 213
Hugging Face
  reference link 339
hybrid filtering approach 376, 377
HyperText Markup Language
  (HTML) 175, 343

# I

identifier (ID) 175
if..else statement 429
if statement 429
image classification/segmentation 377

Immediately Invoked Function
    Expression (IIFE) 26
inheritance 37-40
instances 35
interactive code, writing
    about 68
    CSV files, loading 71
    div container, obtaining for plots 72, 73
    external packages, loading 68-70
    for loop, using 73, 74
interactive computing, Dnotebook
    cell 63, 64
    persistence 66
    state 66
Internet of Things (IoT) 423
iterables 433

# J

JavaScript
    about 423
    conditional branching 428
    data types 425
    fundamentals 424
    loops 428
    machine learning (ML) 337
    variables, declaring 424
JavaScript functions
    about 435, 436
    callback functions 437, 438
JavaScript object
    keys and values 93
jimp
    reference link 337

# K

K-nearest neighbor
    reference link 377

# L

label encoder 142-144
label encoding
    about 142
    reference link 378
layout
    reference link 196
layout margins
    configuring 194, 195
layout size
    configuring 195, 196
let
    versus var 4
line charts
    about 225
    creating, with Danfo.js 225-229
lists 77
local scope
    block scope 11
    function scope 11
logical operations 114-116
logic, on tensors
    reference link 353
loops
    about 428
    do...while loop 432
    for...in loop 433, 434
    for loop 432, 433
    for...of loop 433, 434
    while loop 431

# M

machine learning (ML)
  about 328, 377, 423
  applications 338
  evaluation metrics 333, 334
  inference 330
  in JavaScript 337
  objective functions 332, 333
  problems/tasks 334
  resources 339
  supervised learning 335
  system analogy 328-330
  training 330
  unsupervised learning 336
  working 331, 332
machine learning (ML) tasks
  open source tools 337
map function
  used, for data transformation 131-134
Markdown cells
  about 65
  creating 75
  heading, creating 76
  images, adding 75
  lists 77
  working with 74
Markdown guide
  reference link 76
material-table 273
mathematical computation, on tensors
  reference link 353
matrix operations
  reference link 353
mean encoding 142
mean squared error (MSE) 366
method overriding 38
methods, Danfo.js

arithmetic operations 110-114
filtering 108, 109
loc and iloc indexing 101-104
logical operations 114-116
sorting 105-108
missing values
  replacing 125
  replacing, in DataFrames 127-129
  replacing, in Series 125, 126
Mocha
  about 49
  unit testing 48
modebar
  about 182
  buttons, removing from 183, 184
  configuring 182
  custom buttons, adding to 185, 186
modebar buttons
  reference link 184
modebar, making always visible 182, 183
models, creating with TensorFlow.js
  about 363
  model way 364
  sequential way of 363
mode type
  setting 198
movie recommendation system
  building 380-384
  creating, with saved model 388-390
  embedding layers 385
  embedding layers, merging 385
  inputs 384
  model, building 384-386
  project directory, setting up 381-383
  saving 386-388
  training 386-388
  training dataset, processing 383, 384
  training dataset, retrieving 383, 384

Mozilla Developer Network (MDN)  426
multiple columns
　bar charts, creating from  242, 243
　box plots, making, for  233
　histogram, creating from  238, 239
　violin plots, making, for  234

# N

natural
　reference link  337
natural language processing  377
neural network approach
　used, for creating recommendation
　　system  377
nlp.js
　reference link  337
No-Code Data Analysis/Handling System
　DataTable component  282, 283
　DataTable component,
　　implementing  284-290
　designing  278-282
　environment, setting up  272-277
　file upload  290-293
　laying out  282
　prerequisite tools  272, 273
　state management  290-293
　structuring  278-282
Node.js
　TensorFlow.js, installing  345-348
Node package manager (NPM)  84, 343
Not a Number (NaN)  425
notebooks
　saving  78
number data type  425

# O

object.assign method  23
object data type  427, 428
object elements
　accessing  18
objective functions  332, 333
object-oriented programming (OOP)
　about  34
　classes  35, 36
　inheritance  37-40
objects
　about  18
　cloning  20-23
　copying  20-23
one-dimensional (1D) tensor  351
one-hot encoding
　about  142-146
　reference link  378

# P

padding  192
pandas features
　reference link  83
persistence  66
pipcook
　reference link  337
pixels (px)  192
Plotly.js
　about  174
　bar charts, creating  201-205
　basic charts, creating  196
　box plots, creating  211, 212
　bubble charts, creating  205, 206
　chart types  174

configuration options, for plots 182
data format 178
fundamentals 178
histogram plots, creating 207-211
layout 190
reference link 174
scatter plot, creating 197-201
statistical charts, creating 207
using, via script tag 175-178
violin plots, creating 215-219
Plotly.js, basic charts
reference link 196
Plotly layout
about 190
chart title, configuring 190-192
legends, configuring 192-194
margins, configuring 194, 195
reference link 190
size, configuring 195, 196
Plotly legends
configuring 192-194
plots, configuration options
about 182
default locale, changing 188, 189
download plot options, customizing 187
modebar 182
responsive chart, creating 187
static chart, creating 187
plotting
Danfo.js, setting up for 222, 223
financial dataset, downloading
for 224, 225
Portable Network Graphics (PNG) 187
processed dataset
three-layer regression model,
training with 366-369

promises
about 27-29
used, for cleaning callbacks 29-31
properties
deleting 19, 20
existence, testing 19
Python pandas
reference link 83

## Q

quartiles 211
Query component
implementing 301-303
interface, implementing 303-305
query method
keyword arguments 108

## R

React 83
React-chart-js 273
React.js 272
React-table-v6 273
reader method 119
rechart.js 273
recommendation system
about 374
collaborative filtering approach 375, 376
hybrid filtering approach 376, 377
neural network approach 377
uses 374
reduction operations
applying, on tensors 357
reductions, on tensors
reference link 353

regression model
  building, with TensorFlow.js 357
regression model, building
        with TensorFlow.js
  environment, setting up locally 358
  training dataset, processing 359-362
  training dataset, retrieving 359-362
responsive chart
  creating 187

## S

scatter plot
  about 197
  creating, with Danfo.js 229-231
  creating, with Plotly.js 197-201
scope
  overview 11
script tag, via Plotly.js
  using 175-178
Series
  about 85-92
  bar charts, creating from 241
  box plots, making for 232
  concatenating, along specified
        axis 158, 159
  histogram, creating from 237
  missing values, replacing 125, 126
  violin plots, making for 232
single-column grouped data
  iterating through 252-256
single-column grouping
  about 247-250
  data aggregation 263, 264
six-dimensional (6D) tensor 352
slicing arrays 104

sort_values method
  keyword arguments 105
specific x and y values
  used, for making box plots 235, 236
  used, for making violin plots 236
split-apply-combine 246
spread operator 9, 22
spread syntax
  about 8
  for function arguments 10
  used, for creating new objects
        from existing ones 9
  used, for spreading/unpacking
        iterable into array 8
Stack Overflow
  reference link 83
standardization 261
state 66
static chart
  creating 187
statistical charts
  box plots, creating with Plotly.js 211
  creating, with Plotly.js 207
  histogram plots, creating with
        Plotly.js 207-211
  violin plots, creating with Plotly.js 215
statistical modeling 142
statistics
  calculating 166-168
  calculating, by axis 168-171
string accessors
  about 160-162
  reference link 162
string data type 426
string/text manipulation 159

supervised learning
  about  335, 336
  classification problems  335
  regression problems  335
switch statement  430, 431
syntactic sugar  40

# T

TensorFlow.js
  about  82, 342
  installing  343
  installing, in Node.js  345-348
  layers  342
  reference link  337
  regression model, building with  357
  setting up, in browser  343
  using  343
TensorFlow.js, in browser
  installing, via package
    managers  344, 345
  installing, via script tags  343, 344
TensorFlow.js, in Node.js
  installing, with GPU support  346
  installing, with native C++ bindings  346
tensors
  about  87, 348, 349
  arithmetic operations,
    applying on  354, 355
  basic math operations, applying on  356
  creating  350-352
  operating on  352, 353
  operations on  348, 349
  reduction operations, applying on  357
ternary operator  429
test environment
  setting up  49-54
this property  23-25

this scope  26
three-dimensional (3D) tensor  352
three-dimensional (3D) charts  174
three-layer regression model
  creating  365, 366
  training, with processed dataset  366-369
titanic dataset
  reference link  265
trace  179
trained model
  predict function  369, 370
transfer learning (TL)  342
transpilers
  used, for setting up modern
    JavaScript environment  40
Twitter Analysis dashboard
  backend, building  396
  frontend, building  404
  setting up  394-396
two-dimensional (2D) tensor  351
TypeError  7
TypeScript  423

# U

unit testing
  with Chai  49
  with Mocha  48
unsupervised learning
  about  336
  association problems  336
  clustering problems  336

# V

var
  properties  5, 6
  versus let  4

variables
  accessing, in multiple cells  67, 68
  declaring  424
violin plots
  about  215
  creating, with Danfo.js  231
  creating, with Plotly.js  215-219
  making, for multiple columns  234
  making, for Series  232
  making, with specific x and y values  236
  reference link  219
vocabulary size  384
Vue  83
Vue.js  273

## W

Webpack
  about  41
  reference link  48
  using, example projects  42-48
while loop  431
whiskers  211

## Y

Yarn  343

www.ingramcontent.com/pod-product-compliance
Lightning Source LLC
Chambersburg PA
CBHW081456050326
40690CB00015B/2820